ENVIRONMENTAL SCIENCE, ENGINEERING AND TECHNOLOGY

ORGANOPHOSPHATE PESTICIDES

ENVIRONMENTAL SCIENCE, ENGINEERING AND TECHNOLOGY

Additional books and e-books in this series can be found on Nova's website under the Series tab.

ENVIRONMENTAL SCIENCE, ENGINEERING AND TECHNOLOGY

ORGANOPHOSPHATE PESTICIDES

FABRICE MARQUIS
EDITOR

nova
science publishers
New York

Library of Congress Cataloging-in-Publication Data

ISBN: 978-1-53618-307-8
Names: Marquis, Fabrice, editor.
Title: Organophosphate pesticides / Fabrice Marquis, editor.
Description: New York : Nova Science Publishers, 2020. | Series: Environmental science, engineering and technology | Includes bibliographical references and index. |
Identifiers: LCCN 2020029131 (print) | LCCN 2020029132 (ebook) | ISBN 9781536183078 (paperback) | ISBN 9781536183160 (adobe pdf)
Subjects: LCSH: Pesticides--Environmental aspects. | Organophosphorus compounds--Environmental aspects. | Organophosphorus compounds--Toxicology. | Soil remediation. | Water--Purification.
Classification: LCC TD879.P37 O74 2020 (print) | LCC TD879.P37 (ebook) | DDC 628.5/29--dc23
LC record available at https://lccn.loc.gov/2020029131
LC ebook record available at https://lccn.loc.gov/2020029132

Published by Nova Science Publishers, Inc. † New York

CONTENTS

PREFACE

In this compilation, the authors present progress in research concerning protection from organophosphates activity, as well as an overview of the methods for their removal from the environment and decontamination of contaminated persons.

New developments in research on biosensors is also presented based on the enzyme (or bacteria)-functionalized nanostructures for organophosphorus pesticide detection.

The closing study focuses on how the application of computational methods to the investigation of organophosphorous pesticides is a significant step towards a complete understanding of their behavior in any type of environment.

Chapter 1 - Organophosphates (OPs) are widely used nowadays. They are applied as pesticides, drugs or chemical warfare agents. Their acute toxicity is ascribed to the inhibition of acetylcholinesterase (AChE), a key enzyme in the transmission of nerve impulses in mammals. The mentioned toxic effect is manifested by the accumulation of acetylcholine and can lead to severe neurological disorders, paralysis or death. Besides, chronic exposure to organophosphates is connected with depression and cancer development in mammals. OPs poisoning therapy is based on the application of specific antidotes as well as non-specific and symptomatic procedures. In this contribution, recent achievements concerning the protection from OPs activity and an overview of the methods for their removal from the environment and decontamination of contaminated persons were presented. The new findings regarding the involvement of organophosphates in cancer development are reviewed. Also, attention was paid on the connection between OPs and clinical depression – the link insufficiently investigated and described in the literature.

Chapter 2 - There has been an exponential increase in the usage of organophosphorus (OP) compounds as pesticides. Organophosphorus pesticide analysis has become a progressively important research area due to its extensive application and pollution of the environment. In particular, there is a great need to develop moveable analytical apparatuses that are responsive to remediation and bioremediation process monitoring, where a rapid analysis of a large number of samples is essential. This can be achieved by integrating bio-components with different electrochemical transducers. The close combination of the biological proceedings with the generation of a signal opens the possibility for manufacturing compact and easy-to-use analytical instruments of high sensitivity and specificity. The accessibility of advanced materials, associated with new sensing techniques has led to outstanding innovations in the design and construction of organophosphorus biosensors. Organophosphorus pesticide analysis has become an increasingly significant research area due to its widespread application and contamination of the environment. Distinct from traditional methods of pesticide detection, the application of electrochemical biosensors based on the enzyme (or bacteria)-functionalized nanostructures in the field of pesticide analysis is promising. This review mainly introduces the recent achievements and problems in biosensors based on the enzyme (or bacteria)-functionalized nanostructures for organophosphorus pesticide detection.

Chapter 3 - Application of computational methods to the investigation of organophosphorous pesticides (OPs) represents an inevitable step towards the complete understanding of their behavior in any type of environment. Two major directions are modelling of 1) OP interactions with biomolecules, and 2) their adsorption on inorganic materials. *In silico* investigation of OP interactions with biomolecules is important in the systematic studies of toxicity, for example when the influence of particular functional groups on the overall toxicity is evaluated. Moreover, it is useful when designing potential antidotes and neutralizing agents. On the other hand, investigation of organophosphate interactions with various adsorbent surfaces can provide insight into the efficiency of their removal from the contaminated media - water, soil and air - by adsorption onto various, properly designed substrates. Implementation of *in silico* methods also reduces the exposure of the laboratory staff and equipment to the potential hazards from these highly toxic substances.

Density functional theory (DFT) is a powerful tool for determining of the electronic properties of molecular systems. In investigations of organophosphate pesticides, it has a large-scale application in the estimation of reactivity of the functional groups towards substrates of interest (for example through calculation of Fukui indices). Organophosphate molecule properties, such as geometry, dipole moment, charge distribution, ionization potential and electron affinity, obtained by DFT, are used as descriptors – input parameters for further semiempirical modelling of their interactions with biomolecules. Calculation of adsorption properties on the selected substrates – adsorption geometry, binding strength and charge distribution between the molecule and substrate, are used for the design of innovative, and understanding of existing materials for adsorption, degradation and sensing of organophosphate pesticides.

In: Organophosphate Pesticides
Editor: Fabrice Marquis
ISBN: 978-1-53618-307-8

Chapter 1

ORGANOPHOSPHATES: APPLICATION, EFFECTS ON HUMAN HEALTH AND REMOVAL

Vladan J. Aničijević[1,2,*] ***and Tamara D. Lazarević-Pašti***[3,†]
[1]Faculty of Physical Chemistry, University of Belgrade, Belgrade, Serbia
[2]Ministry of Defence and Serbian armed forces, Belgrade, Serbia,
[3]Vinča Institute of Nuclear Sciences, University of Belgrade, Beograd, Serbia

ABSTRACT

Organophosphates (OPs) are widely used nowadays. They are applied as pesticides, drugs or chemical warfare agents. Their acute toxicity is ascribed to the inhibition of acetylcholinesterase (AChE), a key enzyme in the transmission of nerve impulses in mammals. The mentioned toxic effect is manifested by the accumulation of acetylcholine and can lead to severe neurological disorders, paralysis or death. Besides, chronic

---

* Corresponding Author's Email: lazarevictlj@yahoo.com.
† Corresponding Author's Email: tamara@vin.bg.ac.rs.

exposure to organophosphates is connected with depression and cancer development in mammals. OPs poisoning therapy is based on the application of specific antidotes as well as non-specific and symptomatic procedures. In this contribution, recent achievements concerning the protection from OPs activity and an overview of the methods for their removal from the environment and decontamination of contaminated persons were presented. The new findings regarding the involvement of organophosphates in cancer development are reviewed. Also, attention was paid on the connection between OPs and clinical depression – the link insufficiently investigated and described in the literature.

Keywords: organophosphates, pesticides, nerve agents, toxicity, environment, remediation, cancer, depression

INTRODUCTION

Organophosphates (OPs) are the class of organophosphorous compounds with the general structure corresponding to $O = P(OR)_3$ or $S = P(OR)_3$. They are considered as esters of phosphoric acid. Organophosphates can undergo hydrolysis followed by the release of alcohol from the ester bond. The most important characteristic of organophosphates is the ability to inhibit the activity of the enzyme acetylcholinesterase (AChE). They are mostly known as the main components of different pesticides and warfare agents. However, organophosphates are much more than that. Their remarkable potential as drugs is yet to be discovered. Also, the view on their effect on the human organism is changing. Once considered only as acutely toxic through their influence on AChE, nowadays OPs are in the focus because of the increasing evidence of their carcinogenicity. Moreover, their ability to cause anxiety and depression is newly-observed and still needs to be thoroughly investigated. The aim of this chapter is to review the history of the development of organophosphates, their current status regarding the application, their role in different pathological states, as well as the methods for their removal from the environment and decontamination of exposed persons.

History of the Organophosphates Development

The development of organophosphate chemistry began in the 19th century when Jean Louis Lassaigne synthesized triethyl phosphate in the reaction of ethanol and phosphoric acid [1]. Philippe de Clermont synthesized tetraethyl pyrophosphate (TEPP) in 1854, which was the first OP compound used as an insecticide. Later, the series of highly toxic OP compounds with carbon atom (P-CN) or fluorine atom (P-F) bonded directly to the phosphorus atom, were synthesized [2]. German chemist Gerhard Schräder recognized the potential of TEPP as an insecticide and synthesized the first commercial OP formulation - an insecticide containing this compound as an active ingredient [3].

Although the research has been initiated to obtain new insecticides, it resulted in the discovery of chemical warfare agents (CWA) with nervous-paralytic effects during the time. The first one from the group of these compounds, tabun, was synthesized in 1936, leading to the application of OPs as nerve poisons [3]. The next synthesized CWA were sarin and soman [1]. The synthesis of these compounds as chemical weapons has drawn the attention of the scientific community towards the development of methods for protection against their use in combat operations and potential terroristic activities [1, 4]. After the synthesis of diisopropyl fluorophosphate (DFP), the anti-cholinesterase (anti-ChE) activity of the OP was observed. The term anti-AChE is a common name for all types of acetylcholinesterase inhibitors, which include irreversible, reversible and slow substrates, as well as transition state analogs. For a long time, DFP has been an inevitable model-compound for the study of the pharmacotoxic and toxicodynamic effects as an inhibitor of enzymes from the ChE group. Simultaneously, systemic insecticide tetraisopropyl pyrophosphoramide (iso-OMPA) was synthesized [1, 4]. After the Second World War, research on the synthesis of new CWA was continued and the G series containing GE and cyclosarin were synthesized. Further, new CWA series of OPs known as V-poisons was created, with the most notorious VX poison among them [3, 5].

Applications of Organophosphates

OPs are widely used in agriculture for the protection of plants [2], such as insecticides, herbicides, fungicides, plant growth regulators, rodenticides and as chemosterilants. As mentioned, these compounds also have a role as chemical weapons in chemical warfare, while some of the organic phosphorus compounds are used as medicines [6, 7]. Table 1 provides a list of the most known OPs with chemical formulas and the IUPAC name.

Organophosphate Pesticides

The use of pesticides from the group of OPs is widespread in agricultural production and hygiene of developed countries, where about 150 compounds from this group are currently used [6, 8]. These compounds have different physicochemical properties, such as vapor pressure and solubility in water. They also differ in chemical stability and show diverse toxicity to warm-blooded animals. Depending on the stability in environmental conditions, some compounds are used in plant protection during sowing, while others can be successfully applied only during the harvest period. The first group must be characterized by environmental persistence in order to avoid secondary action and to reduce production costs, while the second group of compounds used in the later period should be less stable. In this way, the concentration of residues is regulated, which must be below a certain threshold during the harvest period. The group of OPs pesticides includes parathion, as the most powerful representative with the widest spectrum of action (due to high toxicity, it is generally no longer applied), methyl parathion, phosphamidon, phosdrin, dichlorophos, chlorthion, thionazin, malathion, dimethoate, chlorpyrifos, etc. An overview of the trade names and chemical formulas of the most commonly used OP pesticides on the market, as well as the physicochemical properties of some of them are given in Table 2.

Table 1. Chemical formulas and IUPAC names of the most known Ops

Trivial name or designation	Molecular formula	IUPAC name
Tetraethylpyrophos-phate, TEPP	$C_8H_{20}O_7P_2$	diethoxyphosphoryl diethyl phosphate
Disopropylfluo-rophosphate, DFP	$C_6H_{14}FO_3P$	2-[fluoro(propan-2-yloxy) phosphoryl]oxypropane
Tetraisopropylpyropho-sphoramide,izo-OMPA	$C_{12}H_{32}N_4O_3P_2$	N-[bis(propan-2-ylamino) phosphoryloxy-(propan-2-ylamino)phosphoryl]propan-2-amine
Acetphate,Orthene	$C_4H_{10}NO_3PS$	N-[methoxy(methylsulfanyl) phosphoryl]acetamide
Azinofos-methyl Guthion	$C_{10}H_{12}N_3O_3PS_2$	3-(dimethoxyphosphino-thioylsulfanylmethyl)-1,2,3- benzotriazin-4-one
Chlorpyrifos, CPF	$C_9H_{11}Cl_3NO_3PS$	diethoxy-sulfanylidene-(3,5,6-trichloropyridin-2-yl)oxy-l-5-phosphane
Diazinon	$C_{12}H_{21}N_2O_3PS$	diethoxy-(6-methyl-2-propan-2-ylpyrimidin-4-yl)oxy-sulfanylidene-$l^{5}-phosphane
Dimethoate, DMT, Cygon	$C_5H_{12}NO_3PS_2$	2-dimethoxyphosphinothioylsulfanyl-N-methylacetamide
Disulfoton, Di-Syston	$C_8H_{19}O_2PS_3$	diethoxy-(2-ethylsulfanylethylsulfanyl)-sulfanylidene-$l^{5}-phosphane
Ethoprophos, Mocap	$C_8H_{19}O_2PS_2$	1-[ethoxy(propylsulfanyl) phosphoryl]sulfanylpropane
Phenamiphos, Nemacur	$C_{13}H_{22}NO_3PS$	N-[ethoxy-(3-methyl-4-methylsulfanylphenoxy)phosphoryl]propan-2-amine
Malathion, Fyfanon	$C_{10}H_{19}O_6PS_2$	diethyl 2-dimethoxyphosphinothioylsulfanylbutanedioate
Metamidophos, Monitor	$C_2H_8NO_2PS$	[amino(methylsulfanyl)phosphoryl]oxymethane
Methidathion, Supracide	$C_6H_{11}N_2O_4PS_3$	3-(dimethoxyphosphinothioylsulfanylmethyl)-5-methoxy-1,3,4-thiadiazol-2-one
Parathion-methyl, Penncap-M	$C_8H_{10}NO_5PS$	dimethoxy-(4-nitrophenoxy)-sulfanylidene-$l^{5}-phosphane
Naled, Dibrom	$C_4H_7Br_2Cl_2O_4P$	(1,2-dibromo-2,2-dichloroethyl) dimethyl phosphate

Table 1. (Continued)

Trivial name or designation	Molecular formula	IUPAC name
Oxidemetone-methyl, MSR	$C_6H_{15}O_4PS_2$	1-dimethoxyphosphorylsulfanyl-2-ethylsulfinylethane
Phorate, Thimet	$C_7H_{17}O_2PS_3$	diethoxy-(ethylsulfanylmethylsulfanyl)-sulfanylidene-$l^{5}-phosphane
Tabun, GA	$C_5H_{11}N_20_2P$	[dimethylamino(ethoxy)phosphoryl]formonitrile
Sarin, GB	$C_4H_{10}FO_2P$	2-[fluoro(methyl)phosphoryl]oxypropane
Soman, GD	$C_7H_{16}FO_2P$	3-[fluoro(methyl)phosphoryl]oxy-2,2-dimethylbutane
Isopropylmethyl pho-sphonofluoridate, GE	$C_4H_{10}FO_2P$	2-[fluoro(methyl)phosphoryl]oxypropane
Cyclosarin, GF	$CH_3P(O)(F)OC_6H_{11}$	[fluoro(methyl)phosphoryl]oxycyclohexane
Amiton, VG	$C_{10}H_{24}NO_3PS$	2-diethoxyphosphorylsulfa-nyl-N,N-diethylethanamine
VX	$C_{11}H_{26}NO_2PS$	N-[2-[ethoxy(methyl)phosphoryl]sulfanylethyl]-N-propan-2-ylpropan-2-amine

Table 2. Physicochemical properties of some organophosphate pesticides

Trivial name	Vapour pressure (mmHg)	Solubility in water (mg dm^{-3})	LC_{t50} inhalation (mg m^{-3})	LD_{50} orally (mg kg^{-1})
	at 20 or 25°C		rat	
Acephate	$1,7 \times 10^{-6}$	$8,18 \times 10^{5}$		700
Azinofos-methyl	$1,6 \times 10^{-6}$	20,9	69	7
Chlorpyrifos	$2,03 \times 10^{-5}$	1,12	>200	82
Diazinon	$9,01 \times 10^{-5}$	40	3 500	66
Dimethoate	$8,25 \times 10^{-6}$	$2,5 \times 10^{4}$		60
Disulfoton	$9,75 \times 10^{-5}$	16,3	200	2,6
Ethoprophos	$3,8 \times 10^{-4}$	750		34
Fenamiphos	1×10^{-6}	329	91	8
Malathion	$3,38 \times 10^{-6}$	143	43,79	290
Methamidophos	$3,53 \times 10^{-5}$	1×10^{6}	162	7,5
Methidathion	$3,37 \times 10^{-6}$	187	50	20
Parathion-methyl	$6,68 \times 10^{-6}$	11	84	2
Naled	2×10^{-4}	1,5		92
Oxidemetone-methyl	$2,85 \times 10^{-5}$	1×10^{6}	1 500	30
Phorate	$6,38 \times 10^{-4}$	50	11	1

Before it is released into the market, the product must be thoroughly examined and identified how dangerous it is for humans and animals [9]. The development of new plant protection products, safe for mammals and fish, includes synthesis, biological screening, soil evaluation, toxicological testing, metabolic studies, compound degradation testing, environmental safety assessment, patent protection and the production of the product itself [10]. In order to bring the environmental impact of pesticides to an acceptable level, all commercially available products are subject to strict legal regulations in most countries [11].

CWA from Organophosphates Group

CWA from the group of OPs are used in military operations to kill, seriously injure or disable exposed persons, expressing their physiological effects [6, 12]. Therefore, chemical weapons, together with nuclear (and radiological) as well as biological weapons (NBC or CBR), are classified as weapons of mass destruction (WMD). In principle, chemical weapons involve agents along with systems for their application. Compared to conventional weapons, a relatively small amount of modern warfare agents from the group of OPs can kill a massive number of people. The most commonly known CWA is sarin, which was also used in terroristic attacks [13, 14].

Research at the molecular level confirmed that CWA induces acute stress in the organism, by almost instantaneous inhibition of AChE and disruption of the concentration of adenocorticotropic hormones in the blood [4]. Taking into account the strength of the toxic effects on the organism and the physiological response of the organism caused by toxic substances, it is possible to make the following classification of CWA: harassing or disabling agents (tear and vomiting agents), incapacitating agents and lethal agents (blister agents, blood poisons, choking agents and nerve poisons).

There are four important series of CWA from OPs group [4, 15]. Series G (G - German) includes tabun (GA), soman (GD), sarin (GB) and cyclosarin (GF). These are high volatility nerve agents that are typically used for a non-persistent to semi-persistent effect. VE, VG, VM, and VX belong to the V series (V - Venomous, poisonous). These agents have low volatility and are typically used for a persistent effect or liquid contact hazard. The third generation of chemical agents is known as persistent. The GV series, the fourth generation of chemical agents, includes GV and Novichok agents. They are designed to have advantages over less stable G-series compounds and highly toxic V-series agents. The researchers also developed agents that belong to a group of quaternary amine salts with improved ability to penetrate neuromuscular synapses. Novichok series includes carbon monoxide phosphorohalides, which also belong to the fourth generation of chemical agents [4, 8]. In addition to OPs, there are some carbamate (CM) nerve poisons. Table 3 provides an overview of CWA compounds belonging to OPs group, together with their basic toxicological and physicochemical parameters and toxicological manifestations in humans.

Table 3. Physicochemical properties, structure and toxicological characteristics of chemical warfare agents (CWAs)

OPs - CWA	GA	GB	GD	VX
X	CN	F	F	$S(CH_2)_2N[CH(CH_3)_2]_2$
R1	CH_2CH_3	$CH(CH_3)_2$	$CH(CH_3)C(CH_3)_3$	CH_2CH_3
R2	$N(CH_3)_2$	CH_3	CH_3	CH_3
Solubility in water (%) at 25°C	10	Mixed	2	3 (∞ < 9.5°C)
Appearance	Colorless to brownish	Transparent, colorless	Transparent, colorless	Clear to yellowish
Persistence in the environment	24-36h	2-24h	Relatively stable	2-6 d
LC_{t50} (mg $min^{-1}m^{-3}$) vapor	200-400	100	50	36-70
IC_{t50} (mg $min^{-1}m^{-3}$) inhalation	100-300	15-75	5-25	5-50
LD_{50} (mg kg^{-1}) intravenous	0,014	0,014	7-12	0,008

ORGANOPHOSPHATES AS DRUGS

Parallel with the discovery of the mechanism of OPs action as AChE inhibitors, the studies on the possible application of diisopropyl fluorophosphate (DFP) in the treatment of gastrointestinal tract, urinary bladder, myasthenia gravis and glaucoma [6, 14, 16]. However, no advantage of OPs regarding the previously used short-acting inhibitors AChE physostigmine and neostigmine were observed. Moreover, the increased risk of cholinergic crises due to overdose, as well as adverse effects on the central nervous system (CNS) due to the increased passage of OP through the blood-brain barrier was observed. Side effects have been partially eliminated during the local application of OPs, so DFP and ecothiopate have been used for several years in the treatment of glaucoma. DFP is also used for the treatment and diagnosis of some forms of glaucoma and other eye diseases, such as accommodation isotropy. Also, some strong poisons, such as paraoxon and armin are now used only in special cases, as the long-term administration has been observed to cause permanent changes in the eye lens [17]. Phosphomycin is used as a broad-spectrum antibiotic [18], and its effect is based on the inhibition of peptidoglycan biosynthesis.

Recent studies have shown that AChE is involved in the development of some types of tumors [19, 20]. These findings have opened the way for the potential use of OPs as cytostatics. Cyclophosphamide or endoxan is used in combination with cis-platinum in the treatment of various types of tumors, as well as autoimmune diseases, where it exerts the dose-dependent effect [21].

Malathion is also used as an ancillary medical device in patients infected with lice and their eggs in the hair. In addition to azinophos methyl, malathion has a high ovicidal activity [22, 23].

ACUTE TOXIC EFFECTS OF ORGANOPHOSPHATES

Due to their extremely frequent use in agriculture, OPs reach the human body through a food chain [24-27]. Their toxic effects are particularly pronounced due to the transition from thio- (P–S) to analogous oxo-forms (P–O), which occurs by biotransformation in the human organism [6, 28-32].

Influence of Organophosphate on Cholinesterases

The common feature of OPs is high toxicity to mammals [6]. By developing new structures of this group of compounds, less toxic OPs were formed, which in turn quickly became the leading class of compounds among insecticides [6]. OPs are specific inhibitors of AChE and butyrylcholinesterase (BChE), whose primary physiological functions are hydrolysis of acetylcholine (ACh) and butyrylcholine (BCh) in cholinergic synapses in the central and peripheral nervous system (Fest and Schmidt 2012). AChE has a very important function in the nervous system. Its primary role is cholinergic activity – ability to catalyze the cleaving of acetylcholine into acetate and choline and also some other choline esters that function as neurotransmitters [6, 7]. This enzyme exerts its cholinergic function mainly on neuromuscular junctions and in chemical synapses of the cholinergic type [6, 7]. The activity of AChE serves to prevent the accumulation of acetylcholine, as this would lead to repeated and uncontrolled muscle stimulation. In this manner, AChE terminates synaptic transmission [6, 7].

The first structure of AChE was determined in the case of the enzyme isolated from electric ray *Torpedo californica* [33]. This result allowed visualization of a binding pocket for acetylcholine at an atomic resolution. Surprisingly, it was shown that the active site of AChE is not on the surface of the protein [33]. It is located at the bottom of a 20 Å deep gorge lined with numerous aromatic residues. AChE isolated from *Torpedo californica* is structurally similar to AChE in vertebrate nerve and muscle [34]. The 3D structure of recombinant human AChE revealed in 2010, confirmed this [35].

First kinetic studies showed that the active site (catalytic anionic site, CAS) of AChE contains two subsites, named "esteratic" and "anionic" subsite [36]. While "esteratic" site is responsible for the catalytic activity of the enzyme, "anionic" site is known to be a choline-binding pocket. Eventually, it was shown that serine and histidine residues located in "esteratic" active site of the enzyme are crucial for the activity of AChE [37, 38]. "Anionic" subsite handles the interaction with a positively charged quaternary group of acetylcholine. Also, "anionic" subsite is the binding site for some quaternary ligands that act as the inhibitors of AChE [39, 40]. In addition, some quaternary oximes, which act as reactivators of AChE after inhibition by organophosphates, bind to "anionic" site.

Besides two subsites of the AChE catalytic center, there is the "peripheral" anionic site (PAS) [41]. It is different from the choline-binding pocket of the active site (Figure 1). This site is involved in substrate inhibition of AChE. However, at low substrate concentrations, its binding to "peripheral" anionic site could accelerate the acylation step in the catalytic pathway [42].

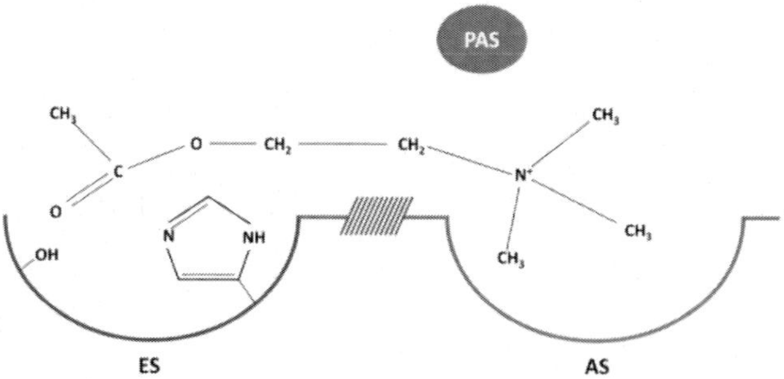

Figure 1. Schematic representation of AChE binding sites: ES – esteratic site, AS – anionic substrate binding site, PAS – peripheral anionic binding site.

Metabolic bioactivation is required for the phosphorylation of enzyme in the presence of compounds containing sulfur which is bonded to a phosphorus atom by a coordinate covalent bond. It is worthily to notice that only compounds with an oxygen atom bound to phosphorus are capable of phosphorylating AChE [43]. Therefore, the bioactivation includes oxidative desulfuration, which is mainly performed in the liver under the influence of cytochrome P-450 enzymes (CYP-450). Although this reaction has been known for decades, the exact CYP-450 isoform that catalyzes oxidative desulfuration is still unknown [8]. In the reactions of biotransformation of OP compounds, the transformation of non-toxic OPs into toxic can occur, as well as also the transformation of one toxic into another toxic compound. There is also the possibility of chemical transformation of toxic OP compounds into compounds that do not have the potency to inhibit AChE.

THE MECHANISM OF ACHE INHIBITION BY ORGANOPHOSPHATES

OPs inhibit AChE activity irreversibly due to the covalently binding to the active site of the enzyme [14, 30]. The first step in the mechanism of AChE activity inhibition in the presence of OPs is the formation of a reversible complex between the acetyl group of enzymes and inhibitor [6, 7]. At the molecular level, the inhibition begins with the formation of an unstable bond between the oxygen of the OP compound and the proton of imidazolium ion, which reduces the electron density of the phosphorus atom. At the same time, the intramolecular hydrogen bond between hydroxyl groups of Ser and Tyr is formed, thereby increasing the electron density on the oxygen atom of Ser. These reactions facilitate the nucleophilic attack on the phosphorus atom. The Ser hydroxyl group reacts with the phosphate ester, leading to alcohol elimination followed by Ser 203 phosphorylation. After this step, the enzyme is irreversibly inhibited. In a further series of reactions, another molecule of alcohol is eliminated.

When OPs molecule contains methyl or ethyl substituents, the spontaneous enzyme regeneration takes several hours. The inhibitory potential of the OPs increases by extending the alkane chain of the substituents, and spontaneous reactivation of the enzyme is negligible. The phosphorylated enzyme cannot hydrolyze ACh, and the inhibition of AChE leads to the accumulation of this neurotransmitter. Thus, the post-synaptic membrane remains depolarized and synaptic transmission does not work. The difference in the behavior of ACh as a substrate and OPs as toxic inhibitors is in the rate of the reactivation of the enzyme. The acyl-enzyme formed in the reaction of the enzyme and ACh rapidly hydrolyzes and regenerates to the free enzyme. The regeneration (deacetylation) of enzymes occurs due to the hydrolysis of the ester bond and is followed by the formation of imidazolium ions. On the other hand, enzyme regeneration after phosphorylation in reaction with OPs is very slow. Phosphorylated AChE is very stable complex, so the activity of enzymes in OPs poisoning without therapy with reactivators cannot be restored. Particularly toxic effects are expressed by the CWA from the OPs group [8, 15].

Symptoms and Manifestation of Organophosphates Toxic Effects

The toxic effects of OPs are the consequence of an excessive accumulation of the AChE substrate – ACh, due to the inhibition of AChE activity. They are manifested by numerous adverse effects on muscarinic and nicotinic receptors in the central or peripheral nervous system. These include Alzheimer's disease, learning difficulty, reduced physical coordination in children, and many other problems listed in Table 4. In extreme cases, these symptoms can lead to death due to paralysis of the respiratory center or respiratory muscles [12].

CWA from the group of OPs, which exert nerve-paralytic effects, belong to the most toxic synthetic compounds which penetrate the body [13]. They have neither smell nor taste and poses significant environmental stability, act very quickly especially when present in high concentrations. Moreover, they do not leave visible traces on the poisoned body and have a cumulative effect that exceeds all known synthetic poisons so far. The lowest limit of irritation is the minimal concentration of CWA needed to induce toxic effects on the body. For example, 10 mg VX induces lethal effects for an adult within 20 minutes after percutaneous contamination, while the lethal dose of sarin due to inhalation works after 2 min.

Table 4. Symptoms manifested in the organism as a consequence of the irreversible inhibition of AChE, by the type of receptor-stimulated and type nerve system (central or peripheral) reacting

Acetylcholine accumulation site	Nicotine and muscarinic receptors and symptoms		
Central neural system	dizziness, headache, anxiety, tremor, confusion, the decline in concentration, mental convulsions, coma and depression of respiratory system, confusion, insomnia, vertigo, sleeplessness, speaking disorders.		
The peripheral autonomic nervous system	Nicotine symptoms	tachycardia, convulsions, muscle weakness, fasciculation and/or paralysis of skeletal smooth muscles, diaphragms and intercostal respiratory muscles, hypertension, uncontrolled muscle contractions and paralysis	
	Muscatine symptoms	*Exocrine glands*	*Smooth muscles*
	Increased elimination (salivation, perspiration, lachrymation), indigestion (spasms, vomiting, diarrhea), lowered heart beat, visual disorders	increased gastric secretion: rhinorrhea, tearing, salivation, sweating, myositis, bradycardia	abdominal cramps, diarrhea, frequent urination, bradycardia, bronchial hypersecretion, bronchospasm, bronchoconstriction, hypersalivation, resulting in incontinence

Previously, it was considered that CWA influence on the body acutely, by inhibiting the enzyme of the nervous system, mainly AChE, whereas low doses do not leave (visible) consequences on the body. However, OPs inhibit also other physiologically important enzymes and thus influence many biochemical processes [30, 44, 45]. There are some examples, such as γ-aminobutyric acid, catecholamines and the inhibition of at least enzymes that participate in physiological processes in the cell (myeloperoxidase, succinate dehydrogenase, sodium-potassium ATPase, etc.). CWA disrupt adenocorticotropic hormone concentrations, and hence all the processes dependent on these hormones (especially in the nervous, reproductive and immune systems). In this way, OPs interfere with all processes in the organism in which inhibited enzymes are involved, such as the production of liver enzymes, regulation of full and adrenocorticotropic hormones, inflammation, and others [44].

ORGANOPHOSPHATES AND CANCER

Cancer is a disease resulting from a combined effect of genetic and external factors acting concurrently and sequentially [46]. Overwhelming evidence indicates that the predominant contributor to many types of cancer is the environment [46]. In accordance with the evidence, The International Agency for Research into Cancer (IARC) classifies carcinogenic substances into four groups (Table 5). For human data, sufficient evidence is defined as the proven causal relationship between exposure to the agent and human cancer. Limited evidence is defined as the observation of a positive association between exposure to the agent and human cancer, for which a causal explanation is considered credible, but probability, predisposition or confounding could not be ruled out with reasonable confidence. Similar definitions relate to the evidence from experimental data [47].

Table 5. Carcinogenicity defined by the IARC [48]

Group	Definition:	Used when:
1	carcinogenic to humans	evidence is sufficient
2a	probably carcinogenic to humans	limited evidence in humans and sufficient evidence in experimental animals
2b	possibly carcinogenic to humans	limited evidence in humans and the absence of sufficient evidence in experimental animals or inadequate or non-existent evidence in humans and sufficient evidence in experimental animals
3	not classifiable as to carcinogenicity to humans	not classifiable to any other group
4	probably not carcinogenic to humans	evidence suggests a lack of carcinogenicity in humans and in experimental animals

As already said, pesticides are extensively used in agricultural, commercial and residential settings, so the exposure of the general population is inevitable. It is hard to avoid the influence of pesticides nowadays. Pesticides and their residues are widespread in almost all parts of the world. They are present in our water, food, and air. Some groups of people, facing pesticides in their workplace, like the workers involved in the production, transportation, preparation and application of these toxins, are even more under the risk due to the exposure to high doses over a long period. Also, many workers are not adequately informed about the risks related to the use of pesticides, particularly in developing countries. That lack of training and equipment to safely handle pesticides increases the health risk for those individuals, but also the general population. Inadequate handling of pesticides leads to their leakage in the environment, resulting in pollution of water and food.

Before, pesticides were considered as acute toxic and not carcinogenic. Today, we know that chronic exposure to pesticides may be a risk factor for microenvironment perturbation and the development of different chronic diseases, ranging from eczema to neurological diseases and cancer [49]. In the past 10 years, an increasing number of case-control and cohort studies, as well as meta-analyses with information on exposure to pesticides and other etiologically relevant factors, have investigated hypotheses connecting occupational pesticide exposure to various types of cancer. Moreover, the evidence is emerging that chronic low-dose exposure to various pesticides perturbs a number of biologic pathways, including oxidative stress [49] and immunotoxicity [49], that have been linked with carcinogenesis. There is evidence that pesticides could be responsible for the development of prostate, ovaries, stomach, bladder, lung, colorectal and pancreatic cancer, as well as for melanoma, neuroblastoma, Burkitt and Non-Hodgkin lymphoma, soft tissue sarcoma, multiple myeloma and leukemia [49].

A recent study aimed to connect changes in levels of different metabolites from urine and saliva due to pesticide exposure with amino acid metabolism, energy metabolism and oxidative stress [50]. These processes are in the base of different diseases including various types of cancer, but also neurological, reproductive and immunological problems, asthma, anxiety, and depression. The mentioned study suggested that oxidative stress due to complex pesticide exposure causes the disturbances in amino acid and energy metabolism, but also could cause other diseases linked to oxidative stress, especially cancer development [50].

Besides the clear involvement of pesticides in cancer development due to oxidative stress enhancement, it has been reported that these toxins can lead to chronic inflammation and increase the production of inflammatory chemokines and cytokines [51]. In a cohort study with greenhouse workers exposed to several neonicotinoids, pyrethroids, and ogranophosphates, the results showed that the levels of IL-22 were significantly increased [51]. This indicated the role of immune deregulation in various pathologies induced by pesticides and cancer development.

The study from India in 2018 revealed the association of oxidative stress with CYP1A2, CYP2B6, CYP2C9, CYP3A4, and PON1 genetic variation in the population exposed to pesticides occupationally [52]. The significantly elevated relative risk of lower antioxidant defense mechanisms (Glutathione, Catalase, Superoxide Dismutase, Glutathione peroxidases, and Glutathione Reductase) in the exposed group was shown [52]. Those data confirmed that exposures to the residues of chlorpyriphos, dichlorvos, ethoprophos, atrazine, butachlor, alachlor, and metolachlor are a major risk factor for increased oxidative stress in the human body [52]. Missense mutations in CYP2B6, CY3A4, and CYP2C9 genes are revealed during genetic variation analysis and have indicated structural and conformation change in protein structure, which could affect their stability and consequently function [52].

Another study in 2018 [53] examined the connection between occupational exposure to a range of pesticides and oxidative stress and DNA damage. Blood samples from 50 farmers exposed to various pesticides for 15 years were tested using the comet assay and the cytokinesis-block micronucleus technique for genetic damage, and the test of thiobarbituric acid reactive substances and catalase activity for the oxidative stress [53]. Cholinesterase activities were also determined. The results showed no significant alteration in cholinesterase activity in comparison to control, but there was a considerable difference in DNA damage and parameters of oxidative stress among groups [53]. The conclusion was that the people exposed to pesticides are more susceptible to genetic damage and, consequently, to diseases resulting from such alterations.

Acute myelocytic leukemia in children is found to be associated with pesticide exposure [54]. Cytochrome P-450 family enzymes are responsible for the transformation of pro-carcinogenic compounds to reactive species which have genotoxic and cytotoxic effects and their activity is affected by the presence of pesticides, such as lindane, parathion, chlorophenol, and atrazine [54].

Among other, organophosphorus pesticides (OPs) are widely used in agriculture are very dangerous and harmful because of their toxic nature [27, 55]. As mentioned, their toxic effects are primarily attributed to the irreversible inhibition of the enzyme acetylcholinesterase [26, 31, 32, 44, 45, 56, 57]. They also inhibit the other enzymes playing an important role in biochemical processes, such as myeloperoxidase (MPO) [44, 45]. Besides the well-known cholinergic function of AChE, there are indications of AChE involvement in oxidative stress, inflammation, apoptosis and cancer development [7]. Acetylcholine level is critical for successful control of inflammation and immune response in peripheral tissues [7]. An increase in acetylcholine level above a certain threshold can suppress the production of pro-inflammatory cytokines [7]. Since AChE is responsible for acetylcholine levels regulation, its role in modulation of inflammation is obvious [7]. Furthermore, the process of inflammation is linked to various conditions [7], including cancer, as already mentioned, so this is one more clue suggesting that AChE is involved in these conditions. It is then logical to assume that organophosphates, as AChE inhibitors, are also important to consider when it comes to cancer development.

Along with the various agents and genetic factors, pesticides have a significant role in colorectal cancer incidence. The study of Abolhassani et al., aimed to investigate the probable involvement of some organophosphorous pesticides and organochlorine pesticides in patients with colorectal cancer [58]. The study showed that a higher level of pesticide residues in patients' serum is associated with higher colorectal cancer incidence. Moreover, it was shown that the increase in serum pesticide residues level significantly stimulated oxidative stress. In general, this research points out a key role of high levels of pesticide residues in serum in colorectal cancer development and their critical role in cancer incidence through food [58].

Malathion is an organophosphate pesticide known for its high toxicity to insects and moderate toxicity to mammals [59]. Besides its acute toxicity, oxidative stress is also reported as a mechanism of toxicity in humans [59]. In the study from 2010 [59], the role of malathion-induced cytotoxicity and genotoxicity is investigated. Due to that, MTT, lipid peroxidation, and single-cell gel electrophoresis assays were performed in order to estimate the levels of cell viability, malondialdehyde production and DNA damage in human liver carcinoma cells, respectively [59]. The results have shown that malathion is mitogenic at lower levels and cytotoxic at higher levels of exposure, and significant lipid peroxidation elevation as well as an increase in DNA damage upon exposure to malathion. All that indicates that exposure to a higher level of malathion induces cytotoxic and genotoxic effects in human liver carcinoma cells and that toxicity is achieved due to oxidative stress [59]. Malathion is also connected to an increased risk of leukemia and lymphoma [60].

Another widely used organophosphate pesticide, chlorpyrifos, is also found to be associated with different types of cancer. The evidence is not exactly consistent, but its involvement in human breast cancer development seems inevitable. The investigation from 2015 [61] tried to identify the pathway involved in chlorpyrifos-inhibited cell proliferation in MCF-7 and MDA-MB-231 breast cancer cell lines, to determine if chlorpyrifos-induced oxidative stress is related to alterations in antioxidant defense system and to elucidate molecular mechanisms underlying in the cell proliferation inhibition produced by the pesticide. It was shown that chlorpyrifos induces redox imbalance altering the antioxidant defense system in breast cancer cells and also that the main mechanism involved in the inhibition of cell proliferation induced by chlorpyrifos directly depends on p-ERK1/2 levels in breast cancer cells [61].

Organophosphates and Depression

Depression is a common mental health condition affecting millions of people around the world. Globally, more than 300 million people of all ages suffer from depression [62]. Many factors have been implicated in the onset and course of the disease including biological, psychological and social factors. The rate of depression seen in women from western industrialized nations is twice that of men [63]. Additionally, individuals who are unemployed, of low socioeconomic status, or who have suffered stressful life events are thought to be at increased risk of developing depression [64]. Disturbances in neurotransmitter functioning and genetic factors have also been implicated in the etiology of depression [65]. Epidemiological studies have also reported higher incidences of depression in particular groups of individuals such as those with chronic health conditions [66] and individuals in specific occupations, such as farming, fishing, and forestry [67]. The reason behind the elevated risk of mood disorder in farming populations is unclear. Some researchers have reported a link between exposure to pesticides, mood disorder and suicidal behavior [68]. Organophosphate pesticides, in particular, are associated with an elevated risk of neuropsychiatric disorder [68]. They are the most widely used group of pesticides in the world and are considered by the World Health Organisation [69] to be one of the most hazardous pesticides to vertebrate animals, responsible for many cases of poisoning worldwide, particularly in developing countries where protective measures are lacking [68]. The neurotoxic effects of high-level acute poisoning are well established and involve changes in peripheral, autonomic and central nervous system function (the cholinergic crisis) resulting in a constellation of physical, cognitive and psychiatric symptoms. However, OPs disrupt many other neurotransmitters and some of these are involved in mood regulation such as serotonin [68]. This could explain the link between pesticide exposure and mood disorder observed in earlier studies. This association appears strongest in individuals who report previous instances of acute poisonings [68]. However, the impact of long- term low-level exposure to OPs (in doses below that causing acute toxicity) on human health is less clear. Some studies have found evidence of ill health, mood disorder and cognitive impairment following low-level exposure to OPs whilst others have not [68].

In the last two decades, only two published studies report evaluations of OP exposed participants using strict diagnostic criteria. First, Amr et al., [70] examined 208 Egyptian pesticide formulators, 172 pesticide applicators, and 223 control subjects. All underwent a full psychiatric interview and diagnoses were made in accordance with the Diagnostic and Statistical Manual of Psychiatric Disorders (DSM-III-R). Psychiatric disorders were found to be more common in exposed subjects, particularly depression and dysthymic disorder. However, as the authors did not collect in-depth information about participants' exposure history, it is impossible to determine whether this is related to long-term low-level exposure, as the possibility of historic acute poisonings cannot be ruled out.

Second, Salvi et al., [71] assessed 37 tobacco workers from Brazil who had been exposed to organophosphate pesticides, evaluating them shortly after exposure and then again, following 3 months without exposure. Diagnoses were made in accordance with the Diagnostic and Statistical Manual of Psychiatric Disorders (DSM-IV). Almost half of the sample was found to be suffering from a psychiatric disorder (mostly anxiety and depression) when first interviewed, but after 3 months of not using pesticides, the rate of mental disorder had dropped by nearly 50% suggesting a strong association between mental health and exposure to pesticides.

Removal and Decontamination of Organophosphates

Removal of Organophosphates from the Environment

The very widespread use of OPs in agriculture contributes to their pronounced presence in food and the environment, which is undesirable due to the toxic effects of these compounds and the products of their transformation [72]. Extreme use of OPs also leads to the accumulation of their residues in the human body through the food chain, and hence to impairing the function of AChE, cholinergic receptor dysfunction, and even death. For these reasons, there has been a need for rapid and effective monitoring, as well as effective removal of OPs, in order to control their levels in food and the environment [73-76]. Today, there are several strategies to eliminate pesticides, which include their adsorption, degradation, and microbiological treatment.

Adsorption

Adsorption of organic pollutants on solid adsorbents is one of the most commonly applied and most effective techniques for removing these compounds from water, as well as detoxification of the organism. For this purpose, natural materials as well as materials derived from organic or industrial waste, which can be classified on carbon, an inorganic oxide, polymeric and similarly, according to their chemical composition. Adsorption of OPs is usually a pseudo-first or second-order process, which follows the model of Langmuir and Freundlich isotherms [77]. Langmuir isotherm suggests monolayer adsorption on a heterogeneous surface with uniformly distributed pores. On the other hand, the diffusion of pesticides in the selected material depends on the physical characteristics and the size of the pores of the solid material. The effectiveness of OPs removal from the environment by various materials depends on the availability of binding sites of adsorbents, as well as on steric disturbances that depend on the structure and functional groups of adsorbates. As an example, one can compare the efficiency of removal of methyl parathion with various solid adsorbents, which, due to steric disturbances and molecular size differences, are higher in seven orders of magnitude compared to ethyl parathion [78].

In the interaction of OPs with carbon materials, the functional groups and the π conjugated system on their surface have a dominant role [79-81]. It was found that π-π interaction on carbon monolithic surfaces leads to the adsorption of pesticides with an aromatic structure, while in the case of pesticides that do not contain an aromatic part adsorption occurs via surface functional groups [81, 82]. In adsorption processes, the most often used adsorbent for OPs removal is activated carbon due to the developed specific surface and porosity, thermal stability and low acid/basic reactivity [15, 83, 84]. Activated carbon from waste tires achieves a level of adsorption of 88.9 mg g^{-1} methyl parathion for 60 min at pH 2 from contaminated water at the initial concentration of 12 mg l^{-1} [78]. In order to use various carbon materials in the adsorption of OPs, special attention has been paid to those obtained by carbonization of waste materials from agriculture, industry or biomass [84, 85].

A large number of literature data related to the testing of pesticide adsorption on mineral materials, zeolites, polymers and other materials [86-90]. Recently, the special interest in mesoporous materials appeared, since they represent a special class of adsorbents with a large internal surface, adjustable size, and pore volume, as well as a stable pore network suitable for modification and functionalization [57, 82]. Mesoporous monetite ($CaHPO_4$) is used as an effective adsorbent for removing malathion from aqueous solutions [88]. The results showed that the adsorption capacity (defined as the mass of malathion per unit mass adsorbent) was 52 mg g^{-1} for the high concentration of malathion, indicating good characteristics taking also into account the low price of this adsorbent. Due to the appropriate characteristics, ordered mesoporous carbons (OMCs) are the promising materials in technologies based on adsorption, in energy and in many other areas of application [57, 82, 91]. A group of carbon mesoporous materials were used to test adsorption of dimethoate and omethoate in order to remove them from aqueous solutions [57]. It has been shown that the adsorption efficiency increases by introducing low concentrations (<1%) of boron, nitrogen or phosphorus into the carbon structure, achieving adsorption capacity up to 164 mg g^{-1} and adsorption of 80% OP, as found for a solution of dimethoate at the concentration of 5×10^{-3} mol dm^{-3}. Recently, the application of nanostructured composites, a biopolymer based on clay minerals with biocompatible materials, demonstrated high efficiency for the effective removal of dichlorvos from the aquatic environment [92-94]. The adsorption efficiency 93.8% for 5 h at 30°C was achieved using MMT-CuO-Ch composite at pH 10.0.

In order to remove OPs from the environment, nanoparticles are also used. It has been shown that silver and gold nanoparticles adsorb chlorpyrifos and malathion, and they are effective and cheap water purification materials when applied onto the aluminum substrate [73]. Dimethoate and its more toxic analog omethoate are also adsorbed onto nanoparticles of spherical or rod-shaped gold [91], which is confirmed by the application of various physical-chemical methods (UV-Vis spectrophotometry, TEM, AFM, and FTIR). The extremely high adsorption capacity for nanoparticles (456 mg g^{-1}) in relation to the rodent nanoparticles (57.1 mg g^{-1}) showed that the adsorption efficiency is highly dependent on the shape and surface coverage of nonmaterial.

Chemical Degradation of Organophosphates

The most important ways of OPs chemical degradation are hydrolysis, oxidation, and photolysis [95-97]. In all ecosystems (soil, water, plants, and animals), degradation begins with the cleavage of P = S bond, and then a more toxic oxo-form (P = O) is formed, replacing sulfur with oxygen in the thiophosphate bond (P = S).

Hydrolysis of OPs strictly depends on the acidity of the environment and the presence of other constituents in solutions or in the soil. If the halftime of hydrolysis in laboratory conditions (pH it is below 6, at a temperature of about 5°C) is 10 days [98], it can increase up to one year in the environment. OPs have been detected in the soil for years after application, which clearly indicates that pesticides can survive for a long time in the environment. It can be explained by their adsorption onto particles in the soil, which makes them inaccessible to microbiological metabolism. Hydrolysis usually begins with the destruction of the phosphorous ester bonds of the thiophosphorous group by nucleophilic or electrophilic attack [99]. In addition to numerous experimental studies of the OPs hydrolysis [95, 99, 100], the *ab initio* test for the alkaline hydrolysis of a group of compounds comprising paraoxon, metal parathion, fenitrothion, demeton-S, acephate, phosalone, azinphos-ethyl, and malathion showed that the conformation of the leaving group affects the mechanism of degradation of OPs thio- and oxo-forms [95].

Chemical degradation of OP is also due to the use of advanced oxidation processes (AOP), which include photochemical (direct photolysis, photochemical oxidation, photocatalysis) [101-104] and ozone-depletion processes. AOPs are based on the production of $OH^{\bullet}$, which indiscriminately attacks most organic molecules, leading to partial or total decomposition and transformation to more toxic and/or stable products [105]. Radiolytic degradation [106] leads to degradation of OPs to less toxic and sometimes biodegradable compounds and often results in the formation of CO_2 and water. By increasing the gamma irradiation dose to 1.0 kGy, the level of diazinon, chlorpyrifos, and phosphamidon in tomatoes decreased by 80–95%, depending on the treated pesticide. AOPs that represent a combination of microwave and UV radiation in the presence of TiO_2 and H_2O_2 allow the removal of up to 100% OP for 6 min from the beginning of the reaction.

Photodegradation under the influence of sunlight is one of the most destructive methods for the degradation of organic compounds [107]. In the case of OPs, this process is of limited importance, because most OPs absorb in the UV region [108]. Several publications dedicated to photodegradation of chlorpyrifos have shown that the combination of UV light with catalysts such as TiO_2, metal ion, and H_2O_2 is very effective [100, 101]. In addition, direct and indirect degradation leads to the formation of various more or less toxic products or to complete mineralization. For chlorpyrifos it has been found that the rate of catalytic photodegradation depends on the nature of water by the order of distilled water > groundwater > lake > river > drinking water [109].

In 2019. Savic et al., [55] investigated the effect of UV-C irradiation, widely used in the food industry, on different formulations containing chlorpyrifos. In this study, technical chlorpyrifos (TCPF) and its oil in water (EW) and emulsifiable concentrate (EC) formulations were irradiated by UV-C, and their photodegradation products were subjected to toxicity assessment, including determination of acetylcholinesterase activity, genotoxicity and oxidative stress using human blood cells as a model system. Toxicity studies were performed using the chlorpyrifos concentrations in the range of those proposed as the maximum residue levels in plant commodities. TCPF, EW and EC photodegradation products induced DNA damage and oxidative stress, and their genotoxicity did not decrease as a function of irradiation time. Irradiated TCPF and EC are more potent AChE inhibitors than irradiated EW. Accordingly, the application of UV-C irradiation was shown not to be a good choice for the decontamination method and must be considered when processing the plants previously treated with chlorpyrifos formulations.

On the other hand, plasma treatment of organophosphate pesticide dimethoate was shown to be successful. Mitrovic et al., reported the method for degradation of dimethoate in water using a non-thermal plasma needle and analyzed the kinetics of dimethoate removal and possible degradation pathways [110]. The effects of dimethoate initial concentration, plasma treatment time, Argon flow rate and the presence of radical promoters on the effectiveness of the proposed method were evaluated. With an argon flow rate of 0.5 slm (standard liters per minute) $1\times10{-}4$M dimethoate can be removed within 30 min of treatment. Using UPLC analysis it was confirmed that one of the decomposition products is dimethoate oxo-analog omethoate, which is, in fact, more toxic than dimethoate. However, the overall toxicity of contaminated water was reduced upon the treatment. The addition of H_2O_2 as a free radical promoter enhances dimethoate removal, while $K_2S_2O_8$ results with selective conversion to omethoate. The feasibility of the proposed method for dimethoate degradation in real water samples was confirmed. The proposed method is demonstrated as a highly effective approach for dimethoate removal without significant accumulation of undesirable toxic products and secondary waste.

Removal of Organophosphates by Microbiological Treatment

One of the promising methods of degradation and detoxification of OPs is enzymatic biodegradation [25, 26, 111, 112]. Several OPs degrading enzymes are known, among which the bacterial enzyme phosphoesterase is best characterized. It has the ability to break down different phosphoester bonds [25, 26, 113]. In laboratory conditions (25°C, pH 7), biodegradation is about the order of magnitude faster than chemical hydrolysis, which is about ten times faster than photolysis [112].

There are a large number of microorganisms in the environment with enzymes capable of digesting OPs. Enzyme oxidation of thio-forms leads usually to the formation of more toxic oxo-forms, which can be used by bioanalytical methods for the detection of low concentrations of these pollutants. Microbiological degradation has a synergistic effect with chemical degradation in the remediation of contaminated water and soil [96].

Some types of bacteria are able to degrade OPs both in water and in the soil. Thus, the biodegradation of chlorpyrifos in the presence of bacterial strain *Pseudomonas kilonensis* (SRK1) isolated from wastewaters with the addition of chlorpyrifos was studied and optimal conditions for the removal of this OP were determined [114]. On the other hand, the bacterial strain, isolated from the sludge collected at the drainage of the chlorpyrifos production plant, degraded the various O-O-dialkylphosphorothioates and O, O-dialkyl- phosphates [115]. This strain, belonging to the genus *Stenotrophomonas*, is able to degrade 100% metal-parathion, metallo-paraoxon, diazinon and phoxime, 95% parathion, 63% chlorpyrifos, 38% of profenofos and 34% of triazophos at a concentration of 50 mg l^{-1} within 24 hours. A new strain of *Bacillus arabhattai* strain, labeled *SanPS1*, has recently been isolated from arable land in India. It is capable to degrade about 56% chlorpyrifos within 24 hours at the concentration up to 500 mg ml^{-1}.

Removal of Organophosphates from Contaminated Faces and Personal Protective Agents

Chemical decontamination aims to neutralize CWA from OPs group within a short time by transforming them into non-toxic products or removing them from the surface, air, water or environment. It is performed by physical, chemical or combined methods. One or more methods of decontamination can be applied as part of the decontamination procedure, in which way the decontamination can be performed by application of chemical, physical or mechanical procedures [116, 117].

The chemical method of decontamination is based on the chemical reactions between decontamination solution and toxic compounds, thereby reducing or eliminating toxicity (harmful effects on biological functions). The physical way of decontamination is based on processes of physical change of the state of toxic substances (evaporation, etc.), whereby they retain the structure and biological activity. The mechanical method of decontamination is based on changing the position of the contaminants, with the toxic substances retaining all their properties. In the chemical methods, the appropriate chemical reactions with the toxic substances must have high reaction rate with large number of CWA, the reaction products should be non-toxic and tolerant for the object that is decontaminated and also for humans (not to irritate the skin, not to be corrosive to metals, and aggressive for clothing). Besides, the reagents (decontamination materials) should be economical for use in large quantities [118, 119].

The choice of the appropriate procedure starts from the type of CWA and the objects that are subjected to decontamination, and also the possibilities of applying the appropriate decontamination type and materials. The following procedures for bringing decontamination materials into contact with CWA are usually applied: dusting, spraying with a solution for decontamination, boiling, immersion in solution and steam (paramedic procedure). The quality of the material for the body protection agents are made is checked by modern devices for measuring penetration and permeation of CWA [119].

Antidotes

The therapy of OPs induced poisoning implies the application of specific antidotes as well as the classical non-specific and symptomatic measures and procedures [120-122]. Irreversibly inhibited enzymes can regenerate their activity only in the presence of AChE reactivators. The most common compounds used to restore AChE activity are those from the oxime group, such as pyridine-2-aldoximes (2-PAM) and 1,1-trimethylene-bis-(4-formyl-pyridinium bromide) dioxides (TMB-4). They can restore the inhibited enzyme activity within a few minutes. However, due to the aging of the pesticide-enzyme complex, the enzyme reactivation becomes impossible [120].

Standard therapy implies the reduction of OPs absorption by stomach due to the administration of activated carbon, accompanied by the administration of an antidote to suppress the effects of the absorbed pesticides [120]. Within the specific therapy, atropine is used as a muscarinic receptor antagonist and oximes as reactivators of inhibited AChE activity. Conventional therapy also implies the use of benzodiazepines, which are effective anticonvulsants, as determined on the basis of experimental results on rodents [123] and primates.

The pharmacological antidote atropine is a competitive AChE antagonist administered intravenously or intramuscularly every 30 minutes, due to its rapid degradation before the signs of atropinization appear (dry mouth, skin, facial flushing, tachycardia, enlarged pupils). Atropine stops the muscarinic effect, while on nicotine it does not work. On the other hand, oximes remove the muscarinic and nicotine effects, while there is no effect on the central nervous system. These compounds reactivate AChE by removing the phosphoryl group. Most commonly used are pyridine oximes such as pralidoxime chloride (2-PAM) or mesylate, P2S (against sarin, cyclosarin, and VX poisons), trimedoxime or TMB-4 and obidoxime or LüH-6 (both against tabun, sarin, and VX- poisons), HI-6 (against sarin, soman, cyclosarin, and VX poisons) and HLö-7 (against all five nerve poisons). Experimental and clinical experiences show that only trimedoxime and obidoxime can reactivate AChE with antidotal protection against most

OPs insecticides [124]. The search for "omnipotent" oxime, effective against all CWA and OP insecticides, is still in progress [125, 126].

Conclusion

Historical representation of organophosphates as pesticides and warfare agents is changing with time. Nowadays, it is shown that they exert tremendous potential as drugs for the treatment of Alzheimer's disease due to their well-known ability to inhibit AChE. Still, they should be used with caution having in mind some new discoveries regarding indications of their involvement in cancer development as well as in anxiety and depression occurrence. Because of all mentioned above, it is necessary to fully reconsider the advantages and drawbacks of organophosphates' use and to establish a new depiction of these intriguing compounds.

Acknowledgments

Authors would like to thank to the Ministry of Education, Science and Technological Development of Republic of Serbia for their financial support (Project No. 172023).

References

[1] Fest, C. and K. Schmidt, *The chemistry of organophosphorus pesticides*. 2012: Springer Science & Business Media.

[2] Kwong, T. C., 2002. Organophosphate pesticides: biochemistry and clinical toxicology. *Ther Drug Monit* 24, 144-149.

[3] Coleman, K., *A History of Chemical Warfare*. 2005: Palgrave Mcmillan UK.

[4] Gupta, R. C., *Handbook of toxicology of chemical warfare agents.* 2015, London: Academic Press.

[5] Montella, I. R., R. Schama, and D. Valle, 2012. The classification of esterases: an important gene family involved in insecticide resistance--a review. *Mem Inst Oswaldo Cruz* 107, 437-449.

[6] Colovic, M. B., et al., 2013. Acetylcholinesterase inhibitors: pharmacology and toxicology. *Current neuropharmacology* 11, 315-335.

[7] Lazarevic-Pasti, T., et al., 2017. Modulators of Acetylcholinesterase Activity: From Alzheimer's Disease to Anti-Cancer Drugs. *Current Medicinal Chemistry* 24, 3283-3309.

[8] Marrs, T., R. Maynard, and F. Sidell, *Chemical warfare agents: toxicology and treatment.* 2007, New York: John Wiley & Sons.

[9] Tang, W., et al., 2016. Independent Prognostic Factors for Acute Organophosphorus Pesticide Poisoning. *Respir Care* 61, 965-970.

[10] Boobis, A. R., et al., 2008. Cumulative risk assessment of pesticide residues in food. *Toxicol Lett* 180, 137-150.

[11] Munoz-Quezada, M. T., et al., 2016. Chronic exposure to organophosphate (OP) pesticides and neuropsychological functioning in farm workers: a review. *Int J Occup Environ Health* 22, 68-79.

[12] Boublik, Y., et al., 2002. Acetylcholinesterase engineering for detection of insecticide residues. *Protein Eng* 15, 43-50.

[13] Tu, A. T., 2007. Toxicological and Chemical Aspects of Sarin Terrorism in Japan in 1994 and 1995. *Toxin Reviews* 26, 231-274.

[14] Vučemilović, A., 2010. Toxicological effects of weapons of mass destruction and noxious agents in modern warfare and terorrism. *Archives of industrial hygiene and toxicology* 61, 247-256.

[15] Gupta, V. K., et al., 2011. Pesticides removal from waste water by activated carbon prepared from waste rubber tire. *Water Res* 45, 4047-4055.

[16] Ballantyne, T. and T. C. Marrs, *Clinical and Experimental Toxicology of Organophosphates and Carbamates.* 1992, Oxford, Boston: Butterworth-Heinemann.

[17] Grant, W. M., *Toxicology of the Eye 3rd ed.* 1986, Springfield. IL: Charles C. Thomas Publisher.

[18] Raz, R., 2012. Fosfomycin: An old--new antibiotic. *Clin Microbiol Infect* 18, 4-7.

[19] Cheng, K., et al., 2008. Acetylcholine release by human colon cancer cells mediates autocrine stimulation of cell proliferation. *Am J Physiol Gastrointest Liver Physiol* 295, 24.

[20] Xi, H.-J., et al., 2015. Role of acetylcholinesterase in lung cancer. *Thoracic cancer* 6, 390-398.

[21] McGuire, W.P., et al., 1996. Cyclophosphamide and cisplatin compared with paclitaxel and cisplatin in patients with stage III and stage IV ovarian cancer. *N Engl J Med* 334, 1-6.

[22] Hoffmann, E. J., S. M. Middleton, and J. C. Wise, 2008. Ovicidal Activity of Organophosphate, Oxadiazine, Neonicotinoid and Insect Growth Regulator Chemistries on Northern Strain Plum Curculio, Conotrachelus nenuphar. *Journal of Insect Science* 8, 29.

[23] Verma, P. and C. Namdeo, 2015. Treatment of pediculosis capitis. *Indian Journal of Dermatology* 60, 238-247.

[24] Kozawa, K., et al., 2009. Toxicity and actual regulation of organophosphate pesticides. *Toxin Reviews* 28, 245-254.

[25] Lazarevic-Pasti, T., et al., 2011. Oxidation of diazinon and malathion by myeloperoxidase. *Pesticide Biochemistry and Physiology* 100, 140-144.

[26] Lazarevic-Pasti, T., B. Nastasijevic, and V. Vasic, 2011. Oxidation of chlorpyrifos, azinphos-methyl and phorate by myeloperoxidase. *Pesticide Biochemistry and Physiology* 101, 220-226.

[27] Lazarevic-Pasti, T. and V. Vasic, 2010. Oxidation of diazinon for the sensitive detection by cholinesterase-based bioanalytical method. *Journal of Environmental Protection and Ecology* 12, 1168-1173.

[28] Colovic, M., et al., 2010. The influence of diazinon and its photoinduced by-products on AChE activity. *Toxicology Letters* 196, S325-S326.

[29] Colovic, M. B., et al., 2011. Single and simultaneous exposure of acetylcholinesterase to diazinon, chlorpyrifos and their

photodegradation products. *Pesticide Biochemistry and Physiology* 100, 16-22.

[30] Colovic, M. B., et al., 2015. *In vitro* evaluation of neurotoxicity potential and oxidative stress responses of diazinon and its degradation products in rat brain synaptosomes. *Toxicol Lett* 233, 29-37.

[31] Lazarevic-Pasti, T. D., et al., 2013. Electrochemical oxidation of diazinon in aqueous solutions via electrogenerated halogens - Diazinon fate and implications for its detection. *Journal of Electroanalytical Chemistry* 692, 40-45.

[32] Lazarevic-Pasti, T. D., et al., 2012. Indirect electrochemical oxidation of organophosphorous pesticides for efficient detection via acetylcholinesterase test. *Pesticide Biochemistry and Physiology* 104, 236-242.

[33] Sussman, J. L., et al., 1991. Atomic structure of acetylcholinesterase from Torpedo californica: a prototypic acetylcholine-binding protein. *Science* 253, 872-879.

[34] Bon, S., M. Vigny, and J. Massoulie, 1979. Asymmetric and globular forms of acetylcholinesterase in mammals and birds. *Proc Natl Acad Sci USA* 76, 2546-2550.

[35] Dvir, H., et al., 2010. Acetylcholinesterase: from 3D structure to function. *Chem Biol Interact* 187, 10-22.

[36] Nachmansohn, D. and I. B. Wilson, 1951. The enzymic hydrolysis and synthesis of acetylcholine. *Adv Enzymol Relat Subj Biochem* 12, 259-339.

[37] MacPhee-Quigley, K., P. Taylor, and S. Taylor, 1985. Primary structures of the catalytic subunits from two molecular forms of acetylcholinesterase. A comparison of NH2-terminal and active center sequences. *J Biol Chem* 260, 12185-12189.

[38] Wilson, I. B. and F. Bergmann, 1950. Acetylcholinesterase. VIII. Dissociation constants of the active groups. *J Biol Chem* 186, 683-692.

[39] Mooser, G. and D. S. Sigman, 1974. Ligand binding properties of acetylcholinesterase determined with fluorescent probes. *Biochemistry* 13, 2299-2307.

[40] Wilson, I. B. and C. Quan, 1958. Acetylcholinesterase studies on molecular complementariness. *Arch Biochem Biophys* 73, 131-143.

[41] Taylor, P. and S. Lappi, 1975. Interaction of fluorescence probes with acetylcholinesterase. Site and specificity of propidium binding. *Biochemistry* 14, 1989-1997.

[42] Johnson, J. L., et al., 2003. Unmasking tandem site interaction in human acetylcholinesterase. Substrate activation with a cationic acetanilide substrate. *Biochemistry* 42, 5438-5452.

[43] Kamel, A., et al., 2009. Oxidation of selected organophosphate pesticides during chlorination of simulated drinking water. *Water Research* 43, 522-534.

[44] Lazarevic-Pasti, T., A. Leskovac, and V. Vasic, 2015. Myeloperoxidase Inhibitors as Potential Drugs. *Current Drug Metabolism* 16, 168-190.

[45] Lazarevic-Pasti, T., et al., 2013. Influence of organophosphorus pesticides on peroxidase and chlorination activity of human myeloperoxidase. *Pesticide Biochemistry and Physiology* 107, 55-60.

[46] WHO, An overview of the evidence on environmental and occupational determinants of cancer, in *International conference on environmental and occupational determinants of cancer: Interventions for primary prevention.* 2011, WHO: Asturias, Spain.

[47] Rushton, L., 2003. How much does the environment contribute to cancer? *Occupational and Environmental Medicine* 60, 150.

[48] IARC, Some Organic Solvents, Resin Monomers and Related Compounds, Pigments and Occupational Exposures in Paint Manufacture and Painting. *IARC Monographs on the evaluation of carcinogenic risks to humans Vol. 47.* 1989, Lyon: IARC.

[49] Alavanja, M. C. and M. R. Bonner, 2012. Occupational pesticide exposures and cancer risk: a review. *J Toxicol Environ Health B Crit Rev* 15, 238-263.

[50] Ch, R., et al., 2019. Saliva and urine metabolic profiling reveals altered amino acid and energy metabolism in male farmers exposed to pesticides in Madhya Pradesh State, India. *Chemosphere* 226, 636-644.

[51] Gangemi, S., et al., 2016. Occupational exposure to pesticides as a possible risk factor for the development of chronic diseases in humans (Review). *Mol Med Rep* 14, 4475-4488.

[52] Kaur, G., N. Dogra, and S. Singh, 2018. Health Risk Assessment of Occupationally Pesticide-Exposed Population of Cancer Prone Area of Punjab. *Toxicol Sci* 165, 157-169.

[53] Hilgert Jacobsen-Pereira, C., et al., 2018. Markers of genotoxicity and oxidative stress in farmers exposed to pesticides. *Ecotoxicology and Environmental Safety* 148, 177-183.

[54] Sabarwal, A., K. Kumar, and R. P. Singh, 2018. Hazardous effects of chemical pesticides on human health Cancer and other associated disorders. *Environmental Toxicology and Pharmacology* 63, 103-114.

[55] Savic, J. Z., et al., 2019. UV-C light irradiation enhances toxic effects of chlorpyrifos and its formulations. *Food Chemistry* 271, 469-478.

[56] Lazarevic-Pasti, T., et al., 2018. The impact of the structure of graphene-based materials on the removal of organophosphorus pesticides from water. *Environmental Science: Nano* 5, 1482-1494.

[57] Lazarevic-Pasti, T. D., et al., 2016. Heteroatom-doped mesoporous carbons as efficient adsorbents for removal of dimethoate and omethoate from water. *RSC Advances* 6, 62128-62139.

[58] Abolhassani, M., et al., 2019. Organochlorine and organophosphorous pesticides may induce colorectal cancer; A case-control study. *Ecotoxicology and Environmental Safety* 178, 168-177.

[59] Moore, P. D., C. G. Yedjou, and P.B. Tchounwou, 2010. Malathion-induced oxidative stress, cytotoxicity, and genotoxicity in human liver carcinoma (HepG2) cells. *Environ Toxicol* 25, 221-226.

[60] Navarrete-Meneses, M. P., et al., 2017. Exposure to the insecticides permethrin and malathion induces leukemia and lymphoma-associated gene aberrations *in vitro*. *Toxicology in Vitro* 44, 17-26.

[61] Ventura, C., et al., 2015. Chlorpyrifos inhibits cell proliferation through ERK1/2 phosphorylation in breast cancer cell lines. *Chemosphere* 120, 343-350.

[62] WHO, *Depression and Other Common Mental Disorders*, WHO, Editor. 2017, WHO.

[63] Patel, V., *Gender in mental health research*. 2005, Geneva: WHO.

[64] Brown, G. W. and T. O. Harris, *Social Origins of Depression: A Study of Psychiatric Disorder in Women*. 1978, London: Tavistock Publications.

[65] Marchand, W. R., D. V. Dilda, and C. R. Jensen, 2005. Neurobiology of mood disorders: a clinical review article. *Hosp. Physician* 43, 17-26.

[66] National Collaborating Centre for Mental, H., *Depression in Adults with a Chronic Physical Health Problem: Treatment and Management*. 2005, British Psychological Society, Leicester (UK).

[67] Sanne, B., et al., 2003. Occupational differences in levels of anxiety and depression: The Hordaland Health Study. *Journal of Occupational and Environmental Medicine* 45, 628-638.

[68] Harrison, V. and S. Mackenzie Ross, 2016. Anxiety and depression following cumulative low-level exposure to organophosphate pesticides. *Environ Res* 151, 528-536.

[69] WHO, 2009. *The WHO Recommended Classification of Pesticides by Hazard and Guidelines to Classification 2009.*

[70] Amr, M. M., Z. S. Halim, and S. S. Moussa, 1997. Psychiatric Disorders among Egyptian Pesticide Applicators and Formulators. *Environmental Research* 73, 193-199.

[71] Salvi, R. M., et al., 2003. Neuropsychiatric evaluation in subjects chronically exposed to organophosphate pesticides. *Toxicological Sciences* 72, 267-271.

[72] Greaves, A. K. and R. J. Letcher, 2017. A Review of Organophosphate Esters in the Environment from Biological Effects to Distribution and Fate. *Bull Environ Contam Toxicol* 98, 2-7.

[73] Bootharaju, M. S. and T. Pradeep, 2012. Understanding the degradation pathway of the pesticide, chlorpyrifos by noble metal nanoparticles. *Langmuir* 28, 2671-2679.

[74] He, J., et al., 2015. Novel restricted access materials combined to molecularly imprinted polymers for selective solid-phase extraction of organophosphorus pesticides from honey. *Food Chemistry* 187, 331-337.

[75] Rasmussen, J.J., et al., 2015. The legacy of pesticide pollution: An overlooked factor in current risk assessments of freshwater systems. *Water Research* 84, 25-32.

[76] Wang, X., et al., 2013. Accumulation, histopathological effects and response of biochemical markers in the spleens and head kidneys of common carp exposed to atrazine and chlorpyrifos. *Food Chem Toxicol* 62, 148-158.

[77] Hameed, B. H., J. M. Salman, and A. L. Ahmad, 2009. Adsorption isotherm and kinetic modeling of 2,4-D pesticide on activated carbon derived from date stones. *J Hazard Mater* 163, 121-126.

[78] Tabassum, N., et al., 2014. Chemodynamics of Methyl Parathion and Ethyl Parathion: Adsorption Models for Sustainable Agriculture. *BioMed Research International* 2014, 8.

[79] Ayranci, E. and N. Hoda, 2005. Adsorption kinetics and isotherms of pesticides onto activated carbon-cloth. *Chemosphere* 60, 1600-1607.

[80] Maliyekkal, S. M., et al., 2013. Graphene: A Reusable Substrate for Unprecedented Adsorption of Pesticides. *Small* 9, 273-283.

[81] Vukcevic, M., et al., 2013. Influence of different carbon monolith preparation parameters on pesticide adsorption. *Journal of Serbian Chemical Society* 78, 1617-1632.

[82] Wu, Z. and D. Zhao, 2011. Ordered mesoporous materials as adsorbents. *Chemical Communications* 47, 3332-3338.

[83] Foo, K. Y. and B. H. Hameed, 2010. Detoxification of pesticide waste via activated carbon adsorption process. *J Hazard Mater* 175, 1-11.

[84] Saleh, T. A. and V. K. Gupta, 2014. Processing methods, characteristics and adsorption behavior of tire derived carbons: a review. *Adv Colloid Interface Sci* 211, 93-101.

[85] Ioannidou, O. A., et al., 2010. Preparation of activated carbons from agricultural residues for pesticide adsorption. *Chemosphere* 80, 1328-1336.

[86] Bruna, F., et al., 2006. Adsorption of pesticides Carbetamide and Metamitron on organohydrotalcite. *Applied Clay Science* 33, 116-124.

[87] Leovac, A., et al., 2015. Sorption of atrazine, alachlor and trifluralin from water onto different geosorbents. *RSC Advances* 5, 8122-8133.

[88] Mirkovic, M. M., et al., 2016. Adsorption of malathion on mesoporous monetite obtained by mechanochemical treatment of brushite. *RSC Advances* 6, 12219-12225.

[89] Valickova, M., J. Derco, and K. Simovicova, 2013. Removal of selected pesticides by adsorption. *Acta Chimica Slovaca* 6, 25-28.

[90] Wang, P., et al., 2016. Preponderant adsorption for chlorpyrifos over atrazine by wheat straw-derived biochar: experimental and theoretical studies. *RSC Advances* 6, 10615-10624.

[91] Momic, T., et al., 2016. Adsorption of Organophosphate Pesticide Dimethoate on Gold Nanospheres and Nanorods. *Journal of Nanomaterials* 2016, 11.

[92] Celis, R., et al., 2012. Montmorillonite-chitosan bionanocomposites as adsorbents of the herbicide clopyralid in aqueous solution and soil/water suspensions. *J Hazard Mater* 210, 67-76.

[93] Pereira, F. A. R., et al., 2013. Chitosan-montmorillonite biocomposite as an adsorbent for copper (II) cations from aqueous solutions. *International journal of biological macromolecules* 61, 471-478.

[94] Sahithya, K., D. Das, and N. Das, 2015. Effective removal of dichlorvos from aqueous solution using biopolymer modified MMT CuO composites: Equilibrium, kinetic and thermodynamic studies. *Journal of Molecular Liquids* 211, 821-830.

[95] Dyguda-Kazimierowicz, E., S. Roszak, and W. A. Sokalski, 2014. Alkaline Hydrolysis of Organophosphorus Pesticides: The Dependence of the Reaction Mechanism on the Incoming Group Conformation. *The Journal of Physical Chemistry B* 118, 7277-7289.

[96] Murillo, R., et al., 2010. Degradation of chlorpyriphos in water by advanced oxidation processes. *Water Supply* 10, 1-6.

[97] Wu, T. and U. Jans, 2006. Nucleophilic substitution reactions of chlorpyrifos-methyl with sulfur species. *Environ Sci Technol* 40, 784-790.

[98] Ellison, D. H., *Handbook of chemical and biological warfare agents.* 2007, London, New York: CRC press.

[99] Lai, K., N. J. Stolowich, and J. R. Wild, 1995. Characterization of P-S bond hydrolysis in organophosphorothioate pesticides by organophosphorus hydrolase. *Arch Biochem Biophys* 318, 59-64.

[100] Hossain, M. S., A. N. M. Fakhruddin, and M. Chowdhury, 2013. Degradation of chlorpyrifos, an organophosphorus insecticide in aqueous solution with gamma irradiation and natural sunlight. *Journal of Environmental Chemical Engineering* 1, 270-274.

[101] Burrows, H. D., et al., 2002. Reaction pathways and mechanisms of photodegradation of pesticides. *J Photochem Photobiol B* 67, 71-108.

[102] Derbalah, A. S., N. Nakatani, and H. Sakugawa, 2004. Photocatalytic removal of fenitrothion in pure and natural waters by photo-Fenton reaction. *Chemosphere* 57, 635-644.

[103] Hirahara, Y., H. Ueno, and K. Nakamuro, 2003. Aqueous photodegradation of fenthion by ultraviolet B irradiation: contribution of singlet oxygen in photodegradation and photochemical hydrolysis. *Water Res* 37, 468-476.

[104] Wu, J., C. Lan, and G. Y. S. Chan, 2009. Organophosphorus pesticide ozonation and formation of oxon intermediates. *Chemosphere* 76, 1308-1314.

[105] Bavcon Kralj, M., et al., 2007. Comparison of photocatalysis and photolysis of malathion, isomalathion, malaoxon, and commercial malathion--products and toxicity studies. *Water Res* 41, 4504-4514.

[106] Ismail, M., et al., 2013. Advanced oxidation for the treatment of chlorpyrifos in aqueous solution. *Chemosphere* 93, 645-651.

[107] Katagi, T., 2004. Photodegradation of pesticides on plant and soil surfaces. *Reviews of environmental contamination and toxicology* 182, 1-189.

[108] Amarathunga, A. A. D. and F. Kazama, 2014. Photodegradation of chlorpyrifos with humic acid-bound suspended matter. *Journal of Hazardous Materials* 280, 671-677.

[109] Muhamad, S. G., 2010. Kinetic studies of catalytic photodegradation of chlorpyrifos insecticide in various natural waters. *Arabian Journal of Chemistry* 3, 127-133.

[110] Mitrovic, T., et al., 2019. Non-thermal plasma needle as an effective tool in dimethoate removal from water. *Journal of Environmental Management* 246, 63-70.

[111] Raushel, F. M., 2002. Bacterial detoxification of organophosphate nerve agents. *Curr Opin Microbiol* 5, 288-295.

[112] Singh, B. K. and A. Walker, 2006. Microbial degradation of organophosphorus compounds. *FEMS Microbiol Rev* 30, 428-471.

[113] Ghanem, E. and F. M. Raushel, 2005. Detoxification of organophosphate nerve agents by bacterial phosphotriesterase. *Toxicol Appl Pharmacol* 207, 459-470.

[114] Khalid, S., I. Hashmi, and S. J. Khan, 2016. Bacterial assisted degradation of chlorpyrifos: The key role of environmental conditions, trace metals and organic solvents. *J Environ Manage* 168, 1-9.

[115] Deng, S., et al., 2015. Rapid biodegradation of organophosphorus pesticides by Stenotrophomonas sp. G1. *J Hazard Mater* 297, 17-24.

[116] Karkalić, R., V. Maslak, and A. Nikolić, 2015. Application of permeable materials for CBRN protective equipment. *Zaštita materijala* 56, 239-242.

[117] Karkalić, R. M., N. D. Ivanković, and D. B. Jovanović, 2016. Dynamic adsorption characteristics of thin layered activated charcoal materials used in chemical protective overgarments. *Indian journal of fibre & textile research* 41, 402-410.

[118] Otrisal, P. and S. Florus, 2013. Application of options of the QCM detection method for the determination of concentrations of toxic compounds depending on resistance assessment of barrier materials. *Hygiena* 58, 125-129.

[119] Otrisal, P. and S. Florus, 2014. Current and perspectives in personal and collective protection against effects of toxic compounds. *Chemicke listy* 108, 1168-1171.

[120] Johnson, M. K., D. Jacobsen, and T. J. Meredith, 2000. Evaluation of antidotes for poisoning by organophosphorus pesticides. *Emergency medicine* 12, 22-37.

[121] Jokanovic, M., 2009. Medical treatment of acute poisoning with organophosphorus and carbamate pesticides. *Toxicol Lett* 190, 107-115.

[122] Kuca, K., et al., 2009. *In vitro* identification of novel acetylcholinesterase reactivators. *Toxin Reviews* 28, 238-244.

[123] Bokonjic, D. and N. Rosic, 1991. Anticonvulsive and protective effects of diazepam and midazolam in rats poisoned by highly toxic organophosphorus compounds. *Arh Hig Rada Toksikol* 42, 359-365.

[124] Stojiljkovic, M. P. and M. Jokanovic, 2006. Pyridinium oximes: rationale for their selection as causal antidotes against organophosphate poisonings and current solutions for auto-injectors. *Arh Hig Rada Toksikol* 57, 435-443.

[125] Murray, D. B., et al., 2012. Rapid and complete bioavailability of antidotes for organophosphorus nerve agent and cyanide poisoning in minipigs after intraosseous administration. *Ann Emerg Med* 60, 424-430.

[126] Peter, J. V., J. L. Moran, and P. Graham, 2006. Oxime therapy and outcomes in human organophosphate poisoning: an evaluation using meta-analytic techniques. *Crit Care Med* 34, 502-510.

In: Organophosphate Pesticides
Editor: Fabrice Marquis
ISBN: 978-1-53618-307-8

Chapter 2

ELECTROCHEMICAL BIOSENSORS FOR ORGANOPHOSPHATE PESTICIDE DETECTION

Nebojša I. Potkonjak*

Department of Chemical Dynamics, Vinča Institute of Nuclear Sciences, University of Belgrade, Beograd, Serbia

ABSTRACT

There has been an exponential increase in the usage of organophosphorus (OP) compounds as pesticides. Organophosphorus pesticide analysis has become a progressively important research area due to its extensive application and pollution of the environment. In particular, there is a great need to develop moveable analytical apparatuses that are responsive to remediation and bioremediation process monitoring, where a rapid analysis of a large number of samples is essential. This can be achieved by integrating bio-components with different electrochemical transducers. The close combination of the biological proceedings with the generation of a signal opens the possibility for manufacturing compact and easy-to-use analytical instruments of high sensitivity and specificity. The

* Corresponding Author's Email: npotonjak@gmail.com, npotkonjak@vin.bg.ac.rs.

accessibility of advanced materials, associated with new sensing techniques has led to outstanding innovations in the design and construction of organophosphorus biosensors. Organophosphorus pesticide analysis has become an increasingly significant research area due to its widespread application and contamination of the environment. Distinct from traditional methods of pesticide detection, the application of electrochemical biosensors based on the enzyme (or bacteria)-functionalized nanostructures in the field of pesticide analysis is promising. This review mainly introduces the recent achievements and problems in biosensors based on the enzyme (or bacteria)-functionalized nanostructures for organophosphorus pesticide detection.

Keywords: organophosphates, pesticides, electrochemical methods, biosensors, enzyme, bacteria, nanostructures

1. Introduction

Pesticides can be defined as chemicals with the capability to eliminate or control the population of the various kinds of pests (rodents, insects, fungi, weeds, etc.). They are categorized as rodenticides, insecticides, fungicides, and herbicides [1]. Pesticides have been extensively used to advance agricultural efficiency by controlling the number of pests and pathogens. Based on its chemical composition, pesticides can be classified into five main groups [1-3]:

- organochlorines,
- organophosphorus,
- carbamates,
- pyrethrin,
- pyrethroids compound

Due to their extensive effectiveness as insecticides, helminthicides, nematocides, fungicides, and herbicides; organophosphorus (OP) compounds have been continuously used as agrochemicals for the pest control [2, 3]. Organophosphorus compounds alone contribute about 38% of total pesticide production and use [1]. Also, beyond 30% of the registered pesticides in the world market and about 45% of those registered with the U. S. Environmental Protection Agency belong to this group of pesticides [4, 5].

However, pesticides show certain toxicity when leached in soil/water/atmosphere, leading to environmental pollution. It is well known that most pesticides have long half-lives and can remain in the environment for decades [6, 7]. Also, these pesticides can easily be transported through rain, run-off or osmosis processes into other aquatic eco-systems [8]. Therefore, the contamination of the environment, and especially aquatic systems caused by pesticides has become a worldwide issue [9, 10].

Among all, the organophosphorus pesticides (OPPs) are the most prevalent insecticides in the world, due to their low cost, simple synthesis and high activity toward insect control. This is way OPP residues can be often found in the atmosphere, soil, groundwater, but also in agricultural products [11, 12]. Even at very low concentrations, these residues can be potentially harmful for the human population and the environment. Precisely, OPP can cause irreversible inhibitory effects on cholinesterase, involved in the transmission of nerve impulses [13]. To prevent OPP from causing such a negative environmental impact, there is an increasing demand to develop sensitive and efficient analytical methods for OPP analysis in food and drinking water [14-16].

2. The Organophosphorus Pesticidesntroduction

Organophosphorus compounds are the chemical compounds that include stable functional groups that contain the carbon phosphorus bond or organic derivatives of inorganic phosphorus acids [17]. These compounds are esters of phosphoric acid with varying combinations of oxygen, nitrogen, carbon, and sulphur attached to them. All the OP compounds shared a common structural pattern [18-20], a phosphorus atom present in the center, which is double-bonded with oxygen or sulphur atom and single bonded with alkoxy/aryloxy/ thioalkoxy groups (R and R') and X any leaving group, Figure 1:

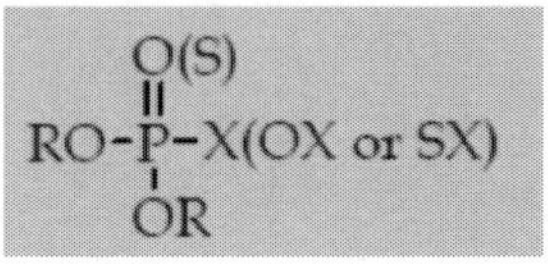

Figure 1. General structure of organophosphorus compounds.

Since organophosphorus compounds displayed low bioaccumulation, high-speed biodegradation, high toxicity, and extensive target range, they are extensively used as pesticides in agriculture for the protection of crops and in domestic for parasitic control in domestic animals and as pest repellent [2, 21, 22]. Based on their structural diversity the most commonly used OP pesticides are chlorpyrifos, paraoxon, malathion, parathion, coumaphos, diazinon, methyl parathion, fenitrothion, and cyanophos. According to their chemical constitution, several types of OP pesticides are identified [23]. Some of the typical OP pesticides are presented in Figure 2.

The biochemical mode of action of OPPs primarily involves the acetylcholinesterase inhibition, occurring throughout the central and peripheral nervous system of vertebrates [24-26]. This involves phosphorylation of the serine hydroxyl moiety of the enzyme active site, thus preventing the hydrolysis of the neurotransmitter acetylcholine, performed equivalently. Subsequently, acetylcholine accumulation at the nerve synapses disturbs the propagation of nerve impulses. Slow recovery of the acetylcholinesterase activity can be observed, because of the spontaneous hydrolysis of the phosphorylated enzyme. The nucleophilic attack of the phosphoryl acetylcholinesterase by reagents (hydroxylamine, oximes) leads to faster enzyme reactivation [24-26].

According to the EPA classification, OPPs are among the most intensely toxic pesticides. They belong to the group of pollutants having toxicity class I (highly toxic) or toxicity class II (moderately toxic). These pesticides are less persistent than the organochlorine pesticides, their extensive practice stances high risk to the whole environment, including the human population. Their pollution consequences from agricultural practices, from industrial waste or discharge, from seepage of buried toxic wastes, and from run-off during spraying [27, 28]. Pesticides production, distribution, usage, exposure, environmental levels, and maximum permissible levels in drinking water and food are a constant subject of various regulations following the national and international legislations. The primarily involved organizations are the: US Environmental Protection Agency (EPA), the EU Commission, the World Health Organization (WHO), the Food and Agricultural Organization of the United Nations (FAO).

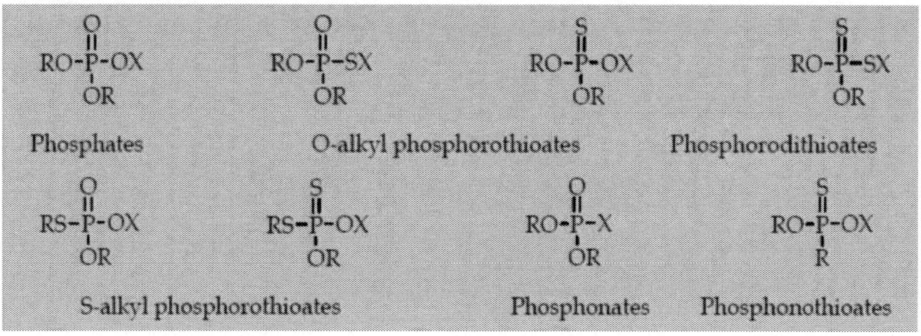

Figure 2. Common types of organophosphorus pesticides (R - usually methyl or ethyl group and X aliphatic, homocyclic or heterocyclic).

Organophosphorus pesticides have a long-term negative disruptive effect on the endocrine system of fish, birds, and mammals [29, 30]. Though numerous research it was confirmed the capability of pesticides to cause endocrine disturbance due to their interaction with thyroid, androgen, estrogenic receptors and other parts of the endocrine system, many long-term effects are still unidentified [31-34]. Usually, pesticides are realised in the environment by various physical, chemical and biological processes. To prevent problems of pesticide contamination, the amount of OPPs pollution must be evaluated. Besides human exposure, there is also an alarm that organophosphorus pesticides could leak into the ground and public water supplies and pollute the surrounding environment. Reports in the literature have shown concern over exposure to non-target organisms such as birds and fish, as well as the potential for human exposure from sources such as fresh fruits and vegetables and processed foods [35, 36]. These neurotoxic compounds, which are structurally similar to the nerve gases, irreversibly inhibit the enzyme acetylcholinesterase, essential for the functioning of the central nervous system in humans and insects, resulting in the build-up of the neurotransmitter acetylcholine which interferes with muscular responses and produce serious symptoms in vital organs and finally death [35, 36].

Generally, chromatographic methods such as gas chromatography, liquid chromatography, and mass spectrometry are implied in sensing of the presence of pesticides in environmental samples. Nevertheless, these methods need strict detection conditions, making them unsuitable to deal with a large number of samples. Consequently, new methods which are low cost and high efficiency are necessary. An operative approach for detecting OPPs must be based on utilizing enzymes since numerous pesticides are designed to inhibit diverse enzymes within pests [37]. Looking from this perspective, enzymes such as acetylcholinesterase (AChE), butyrylcholinesterase (BChE), alkaline and acid phosphatase (ACP), tyrosinase, organophosphorus hydrolase, aldehyde dehydrogenase, and others can be employed to sense pesticides in soil and aquatic systems, and even in beverages and food [38, 39]. Presently, biosensors based on enzyme-functionalized or bacteria-functionalized nanostructures exhibit efficiency and practicality, attracting great attention [40, 41].

3. The Electrochemical Biosensors

Biosensors are the analytical devices, which are capable to detect target analyte of particular concern. All biosensor involves following [42]:

- bio-recognition element, which distinguishes the physical, chemical or biological reply from the analyte,
- transducer, which alters generated reply into the computable signal.

These biosensors deliver elucidations for analytical measurement of any kind of analyte in both laboratory and field testing. Consequently, biosensors are important for inclusive effectiveness in safety measurements of food, monitoring of environmental issues, as biological and chemical warfare agents, agricultural product safety, clinical diagnostics and biomedical research to process control.

The electrochemical biosensors process is based on the usage of a system biological component/bio-receptor engaged in straight connexion with an electrochemically active transducer (electrode) to get an analytical signal by coupling bioelectrochemical interactions [43, 44]. The principle of electrochemical sensors can be explained as follows: an electroactive analyte is subjected to fixed or varying potential causes oxidation or reduction of analyte on the working electrode surface, which leads to the generation of an electrochemically measurable signal by the variation on electron fluxes (electric current). This signal can be measured by the electrochemical device [45-50].

Being a product of the present development in biotechnology and material science, biosensors as analytical devices follows the contemporary policies of the chemical information transduction. Due to their selectivity and specificity biosensors are considered promising candidates for sensing OPPs. The electrochemical biosensors use a biological recognition element retained in direct spatial contact with an electrochemical transducer to obtain an analytically useful signal by coupling biochemical and electrochemical interactions [43, 51]. The biorecognition elements or bio-receptors, according to the biochemical event, can be divided into two main types: biocatalytic and bio-complexing (bio-affinity based) [43]. The biocatalytic ones include enzymes, whole cells (bacteria, fungi, eukaryotic cells or yeasts) or cell organelles and particles (mitochondria, cell walls), and tissues (plant or animal tissues). Regarding the bio-complexing receptors, antibodies, biomimetic materials, cell receptors, and nucleic acids are mainly engaged.

The applied electrochemical transduction mode [43] can be obtained by using potentiometric and amperometric methods which can be further divided as static and dynamics electrochemical methods [51, 52]. The potentiometric determination is founded on the measurement of the electromotive force (*EMF*) of a galvanic element, constituted of an indicator (working) electrode and a reference electrode (two-electrode electrolytic system). The potential (E) of the working electrode depends on the concentration of an analyte (according to the Nernst equation) while the potential of the reference electrode is always constant during the measurements. The exponential character of the relationship between the potential of the indicator electrode and the analyte concentration defines the wide concentration range for determination (from 3 to 4 decades), but also the low accuracy and precision of the given method [43, 47].

The amperometry involves the measurement of the current response recorder by an indicator electrode which can be recorded potentiostatically or potentiodynamically [45-50]. The electric current response is proportional to the concentration of the electroactive species. There are several advantages of the amperometric methods [43]:

a) The process is controlled by the applied electrode potential;
b) Sensitivity and precision of the determinations is high;
c) The calibration plot is always linear.

The maintenance of the biological constituent of the electrochemical biosensor in connection with the transducer is achieved by its immobilization. Since the biorecognition element is usually an enzyme, the term enzyme immobilization is mostly used. It labels enzymes physically restricted at or located in a definite region or space with preservation of their catalytic activity and which can be used repetitively and continually [53, 54]. This method guarantees some subjects: the effective use of the enzyme and its stabilization, the localization of the interaction, the prevention of product contamination, etc. Biorecognition elements immobilization is reached applying numerous methods including membrane entrapment, polymeric matrix, self-assembled monolayers entrapment, covalent bonding, and bulk modification of electrode material (carbon paste or graphite epoxy-resin). The biosensors are usually self-contained, simple to handle and capable to deliver data in real-time, without or with minimum sample preparation. These performances, in concert with their sensitivity, selectivity, and low cost, make them suitable for "in-field" and "on-line" analysis, and an excellent complement to the expensive and time-consuming classical analytical techniques [55, 56]

4. Enzime Electrochemical OP Biosensirshe

According to the nature of the biological recognition element, the electrochemical biosensors for OP pesticides analysis could be classed into two major groups [43, 57]:

- enzyme electrochemical OP biosensors
- bacteria electrochemical OP biosensors

Enzymatic biosensors employ enzymes (either inhibiting or enzymatic biosensors employ enzymes (either inhibiting or hydrolytic) near the transducer element facilitating high and specific reactivity with their substrate, at an ease. In OP electrochemical biosensors, the target analyte can be detected employing [57]:

1) an inhibition mechanism, where, the OP pesticide inhibits the enzymatic activity of enzymes like AChE, BChE, tyrosinase and alkaline phosphatase and thus measuring the decrease in its response,
2) a catalytic mechanism, where the hydrolytic ability of enzyme, OPH to catalyzes the hydrolysis of OP compounds is directly measured, which is increased in accordance to OP compound. Thus, in the field of bio-sensing for OP compounds, enzymatic biosensors represent the most attractive area of research due to their fast response, high robustness, and easiness of immobilization [57].

5. Electrochemical OP Biosensors Based on Enyime Inhibition Mechanism

Elementary principles of these OP biosensors are that in the existence of the OP compounds, neural toxins and other drugs, enzyme AChE or BChE is incapable to change the substrate acetyl thio-choline or butyryl thiocholine into thiocholine and acetic acid/butyric acid. Thio-choline if produced (in absence of OP compounds) undertakes oxidation/dimerization under applied potential, but in presence of OP compounds, the electric current produced due to oxidation at anode has opposite relation to the number of OP pesticides in the certain sample [58]. Other enzymes like tyrosinase, a copper-containing enzyme, which oxidizes easily the phenols (such as tyrosine) to o-quinones and a Mg^{2+} and Zn^{2+} holding alkaline phosphatase enzyme recognized to catalyze the hydrolysis of phosphorylated compounds with wide substrate specificity, get inhibited owing to the presence of various drugs, inorganic salts, heavy metals, toxic environmental pollutants including carbamates and OP pesticides. The inhibition in the activity of these enzymes results in the decrease of biosensor response measurement and can be quantified as the number of OP compounds present in the given sample.

The enzyme inhibition based OP biosensors can be classified based on various matrices used for immobilization of enzymes, on the physical and chemical interactions, between them. Based on of the transducer used, enzyme inhibition based OP biosensor can be classified into four types [57]:

1) electrochemical (amperometric and potentiometric)
2) thermal
3) piezoelectric
4) optical (fluorescence and surface plasmon resonance).

The enzyme inhibition-based biosensors for the determination of OP pesticides is described by the following mechanism:

$$\underset{(\text{Enzyme})}{\text{EH}} + \underset{(\text{OP pesticide})}{(\text{OR})_2\overset{\overset{\text{O}}{\|}}{\text{P}}\text{-X}} \rightarrow \underset{(\text{Phosphorylated enzyme})}{(\text{RO})_2\overset{\overset{\text{O}}{\|}}{\text{P}}\text{-E}} + \text{HX}$$

Electrochemical OP biosensors evaluate the electroactive species generated by the biological component or bio-receptor (enzymes – AChE,

BChE, tyrosinase, and alkaline phosphatase) with the assistance of electrochemically active transducer element i.e., electrode. The signals produced so can be easily monitored by either change in the current generated by redox reaction as in amperometric biosensors or by a change in the pH of the reaction medium as in potentiometric biosensors [59].

In enzyme-based amperometric biosensors, net current is produced owing to the catalyzed redox reactions of biological recognition element (enzymes) under a constant electrode potential that is applied between a working electrode and a reference electrode and the size of this current is directly proportional to the concentration electroactive species which is present in the sample. AChE activity is inhibited by the occurrence of OP pesticides and by the usage of different kinds of immobilization matrices for AChE, various amperometric biosensors with different detection limits and linearity range for OP pesticides have been fabricated so far [60]. Among the large number of fabricated amperometric AChE based OP biosensors, the lowest detection limit (1×10^{-11} μM) and linearity range (1.0×10^{-11} – 1.0×10^{-2} μM) was detected by the biosensor having immobilized AChE on o-phenylenediamine/carbon/cobalt pthalocyanine (CoPc) screen-printed electrode for detection of OP compounds (dichlorvos, parathion and azinphos) [61]. The enzyme was entrapped in the immobilization matrix and showed good storage stability of 92 days with an incubation time of 10 min. Nevertheless, AChE immobilized onto cellophane membrane/Au electrode via crosslinking showed the highest limit of detection (LOD) (1.45 μM) and linearity range (1.45 – 7.26 μM) with 15 min of incubation time for detection of paraoxon [61]. To detect paraoxon in water samples and standard pesticide samples, amperometric biosensors were fabricated by entrapping BChE alone and BChE in combination with choline oxidase into kappa a kappa-carrageenan gel. The biosensors showed a LOD of 4.5 μg L^{-1} and 4.8 μg L^{-1} respectively [62, 63]. The amperometric biosensor having BChE immobilized over the Prussian blue modified screen-printed electrodes showed a LOD of 4 ppb for paraoxon and 1 ppb for chlorpyrifos-methyl oxon [61]. In another work, it was proposed a BChE based amperometric biosensor embedded in a flow system by cross-linking BChE onto a screen-printed electrode having Prussian blue nanoparticles [64]. The LOD for the detection of paraoxon was 1 ppb with storage stability of 60 days. BChE tends to show rather higher storage stability and hence could be used easily for future commercialization in the agro-food biosensor market. In another amperometric study, biosensor fabricated with tyrosinase cross-linked onto CoPc showed a linearity range of 2.28×10^{-8} – 3.8×10^{-7} M for methyl parathion and 6.24×10^{-8} – 1.64×10^{-7} M for diazinon [65]. However, tyrosinase entrapped and cross-linked to 1,2-naphthoquinone-4-sulfonate (NQS) and Prussian blue revealed a LOD of 6×10^{-8} M and 10^{-7} M and linearity up to 8×10^{-6} M and 10^{-7} – 10^{-6} M for dichlorvos and paraoxon OP pesticides respectively [66, 67]. A novel style of organic phase enzyme sensors having immobilized tyrosinase onto the carrageenan gel was utilized for the detection of dimethoate, paraoxon and malathion OP pesticides at a LOD of 10^{-6} M, 5×10^{-6} M, 5×10^{-6} M and linearity range of 2×10^{-6}–0.2 M, 10^{-5}–10^{-2} M, 10^{-5}–10^{-2} M [68]. To detect OP compound, malathion amperometric biosensor based on immobilization of the enzyme alkaline phosphatase obtained from thylakoids of spinach via a new method called as a laser-induced forward transfer for achieving fast and reproducible detection with a lower value of 0.001 ppb and linearity range of 0.2–45.0 μg L^{-1} [69]. In another biosensor, alkaline phosphatase cross-linked with algae-bovine serum albumin (BSA)/ZnO nanoparticles/glassy carbon electrode was used for the voltammetric detection of chlorpyrifos without any kind of interferences from other pesticides (acephate, malathion, triazophos) as well as alkali metals [70]. These inhibition-based amperometric OP biosensors showed a rapid response and high sensitivity with quite a wide linear range and LOD. Low specificity due to interfering substances such as heavy metals, carbamate pesticides and long incubation steps were the major limitations for these inhibition based biosensors.

Basic principles of the potentiometric biosensors depend on the capability of conversion of hydrogen (H^+) ions either generated or absorbed by ion-selective electrodes (transducing element) into an electrical signal. A change in pH directly depends on the target analyte present in the given sample [55]. To detect dichlorvos, a potentiometric biosensor was fabricated by covalently linked AChE on PEI-coated GCE, which showed a LOD of 1.0 μM with an incubation of 10 min [71]. AChE was also cross-linked with glutaraldehyde onto nylon and cellulose nitrate membrane/pH electrode, which exhibited LOD of 0.038 μM and linearity range of $50\times10^3 - 2.5\times10^3$ μM for trichlorfon detection at an incubation time of 15 min and good storage stability of 30 days [72]. A potentiometric BChE inhibition based OP biosensor having entrapped BChE onto plasticized polyvinyl chloride in the pH range of 2.0–10.5 showed an improved selectivity with linearity range of 10^{-6}–10^{-8} M for profenofos and $10^{-5} - 10^{-6}$ M for 2, 2 dichloro vinyl dimethyl phosphate (DDVP) [73]. The BChE based potentiometric biosensor with highly sensitive disposable screen-printed heptakis(2,3,6-tri-o-methyl)-β-cyclodextrin (β-CD) as ionophore was constructed for the detection of malathion OP pesticide in human serum with a LOD of 8×10^{-7} mol L^{-1} and linear range from 10^{-6} to 10^{-2} mol L^{-1} [74]. The LOD and enhanced reproducibility achieved by these potentiometric OP biosensors were adequate to monitor contaminations in food and other agro-products. It appears that the requirement of high potential and fouling problems affects the detection of thiol and OP compounds.

Carbon nanotubes (CNTs) are novel nanomaterials developed lately, which have excellent performance in chemical and biological sensing applications [75]. They are hollow graphitic cylinders with fast the electron-transfer rate and the electrocatalytic effect [76]. CNTs have many outstanding properties, such as high chemical stability, unique electronic properties, very high mechanical difficulty and strength [77]. Thus CNTs have excellent prospects in future applications, particularly in field-effect transistors, nanoprobes, bioelectronics, and sensors [78-80]. CNT-modified electrodes display excellent electrocatalytic activity for the reaction of hydrogen peroxide and nicotinamide adenine dinucleotide (NAD) due to the rapid electron transfer capability of CNTs. It was observed that the overvoltage of hydrogen peroxide and NAD decreased sharply and the sensitivity increased, showing that CNT-modified electrodes have great prospects for use in dehydrogenase and oxidase-based amperometric biosensors. Specifically, CNT hybrid materials have captivated researchers' attention due to their promising applications in highly sensitive biosensors and nanoelectronics. Over the past decade, there has been some progress in carbon nanomaterials. The best amperometric reaction of $CNT\text{-}NH_2$ was obtained on the AChE/CNT-NH_2/GC electrode [81]. This offers some future predictions related to fabricating new amperometric biosensors for the detection of OPs through surface functionalization, based on the assembly of nanomaterials onto liposome bioreactors. A novel amperometric biosensor for detection of OPs shows successfully immobilizing AChE by covalent linkage onto a modified graphite electrode, and functionalizing CNTs by electrochemical treatment [85].

Gold nanoparticles (AuNPs) are the most stable metal nanoparticles. AuNPs have captivated the attention of researchers due to their unique characteristics such as the behavior of the individual particles, optical properties (quantum size effect), size-related electronic properties, magnetic properties, and applications to catalysis and biology [83, 84]. Recently, AuNPs have been confirmed to be very valuable in catalytic applications even at low temperatures, even though gold is well known to be inert, and the fields of AuNPs-catalyzed CO reduction, O_2 reduction and methanol oxidation [85, 86].

Silver nanoparticles (AgNPs) have many exceptional properties, like large surface area, excellent catalytic activity, high conductivity, and biocompatibility. Furthermore, AgNPs can effectively catalyze hydrogen peroxide reduction and can hold their biological activity for biomolecule immobilization in an appropriate microenvironment [87, 88]. The preparation of electrochemical biosensors with improved analytical performance using AgNPs has been developed, based on the capability of AgNPs to facilitate more efficient electron transfer between the immobilized biomolecules and electrode substrates. This provided an effective and simple platform for enzyme immobilization on the electrode surface. Similar to this, a novel, simple and rapid method of analysis for OPs was proposed [89]. Namely, the method uses thiocholine, which is designed by AChE-catalyzed collapse of ACh, as an aggregator of AgNPs. With the research into the development of biosensors, the joint application of various materials has been found to bring more acceptable results. AgNPs have always been a hot spot of research due to their outstanding features, but have not played a major role in the study of enzyme-based biosensors to detect OPs. This can be attributed to the preparation of AgNPs. Although the preparation methods of AgNPs are mature, there are still problems with controlling the size uniformity and crystal shape stability of AgNPs, improving their morphology and avoiding agglomeration. However, with the in-depth study of the properties of AgNPs and the continuous development of their application perspectives, AgNPs will play a more important role in many fields.

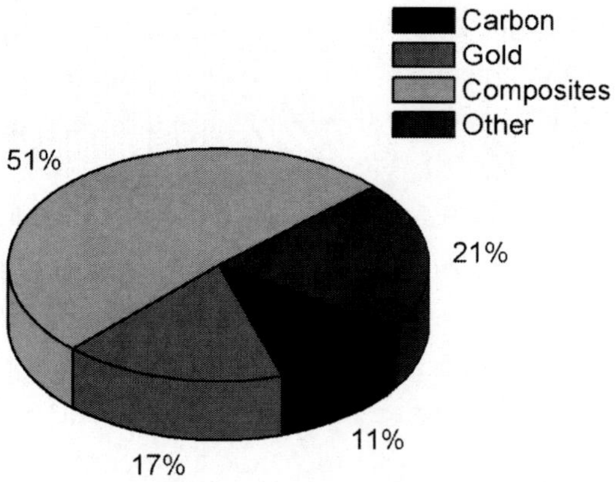

Figure 3. Studies of organophosphorus detection by enzyme-based biosensors with nanostructures functionalized using different nanomaterials.

Zirconia (ZrO_2) has many exceptional properties and a wide range of applications. It has already become one of the most industrially significant materials. Conventionally, ZrO_2 was applied for foundry sands and flours, refractory ceramics, and abrasion-resistant materials. With continuous technological development, other applications of ZrO_2 have been found including oxygen sensors, catalysts, resistive heating elements, fuel cells, and jewelry. In the following explanation, the main focus has been given on the aspect of its application to biosensors. Researchers have exploited zirconia for making numerous films by self-assembly, and have developed a DNA biosensor by using a zirconia-based DNA probe, based on the strong affinity of zirconia for the phosphoric group [90]. Some research has reported the preparation of zirconia films or microcrystals by electrodeposition of $ZrOCl_2$ at bare or functionalized gold surfaces. Proposed a novel method utilizing a gold electrode modified with zirconia nanoparticles to achieve electrochemical sensing of nitroaromatic OPs [90]. In recent research, a novel method was proposed that used self-assembled monolayer (SAMs) to electrodeposit a film of zirconia nanoparticles (ZrNPs) on a gold electrode. Due to the highly stable and selective adsorption of molecules containing phosphoric groups onto the ZrNPs film, a biosensor was developed for the determination of parathion, based on its interaction with the ZrNPs electrodeposited on the modified gold electrode [91]. Still, the surface energy of zirconia nanoparticles is high, and they can easily be reunited and form a single particle. Thus, it is problematic to achieve nanoparticles with good dispersibility and dispersion stability. This makes it difficult to achieve large surface area effects, volume effects and quantum size effects for these nanoparticles. Efforts to tackle this problem have involved the modification of nano-zirconia and its composite powder, which has increased its dispersion stability and reduced the agglomeration of particles. This is broadening the field of its application.

Cadmium sulfide (CdS) is an important semiconductor material with a band-gap of 2.4 eV and unique photochemical properties. It is widely used in photochemical batteries and energy storage devices. CdS nanomaterials have extensive applications in many fields, such as photoluminescence, electroluminescence, sensors, infrared window materials, and photocatalysis. However, the performance is dependent on grain size. The research into CdS nanomaterials has attracted widespread attention.

At present, the main problem with CdS nanomaterials is the various limitations of the preparation methods, such as the harsh reaction conditions, the great influence on the environmental atmosphere, the poor retention of the samples over long periods, the poor crystallinity, the difficulty of obtaining the materials in powder form, and others. The present synthesis methods suffer from the drawback of large surface area, high surface free energy, easy agglomeration and poor stability of the particles. Furthermore, the compatibility of CdS with polymers is poor, and it is difficult to evenly disperse it in the polymer matrix. Consequently, the surface variation of CdS nanomaterials needs further development. Also, the preparation of CdS nanomaterials is mostly at the laboratory stage, and more research still needs to be done in industrial production. The application of CdS nanomaterials' luminescence is mostly limited to the study of luminescence intensity and fluorescence. Other aspects need to be further studied, such as the CdS luminescence mechanism, multi-component co-doping, and various composite materials. Other applications should be further developed. Through a deeper understanding of these issues, CdS nanomaterials can be more widely applied.

With the progress of in-depth research, the drawbacks of single materials in research and practical application are gradually emerging, while composite nanomaterials exhibit superior properties. For example, the combination of graphene and some metals offers many advantages. Graphene has fascinated many researchers for its novel physical properties, two-dimensional structure and potential applications in transparent conductors and nano-electronic devices since it was discovered in 2004. The many excellent properties of graphene include low cost, large surface area and excellent electrical conductivity, making it an ideal material for electrochemistry. Some metals, when introduced into graphene, strengthen its activity, including Au, TiO and SnO_2. Since these kinds of nanocomposite films can generate synergistic effects to enhance the sensitivity of graphene, graphene-based nanocomposites have developed as an enhanced sensing platform for biosensors. A simple method to efficiently prepare Prussian blue nanocubes/reduced graphene oxide was introduced. Subsequently, they developed a novel AChE biosensor based on graphene oxide (GO) in polyethyleneimine aqueous solution [92]. A novel sensor based on the combination of single-walled carbon nanotubes (SWCNTs) and hybrid nanocomposites consisting of copper oxide nanowires (CuO NWs) was proposed for the detection of OPs. It demonstrates that the newly developed CuO NWs–SWCNTs hybrid nanocomposites can be applied for simple, fast, selective and sensitive analysis of OPs. Nevertheless, this process is very complex, which decreases operability. As a challenge, this method remains to be improved in the future. The method developed by El-Moghazy and co-workers is found to be much simpler [93]. They proposed a novel detection method for phosmet in olive oil, based on a genetically-engineered AChE immobilized in an azide-unit water-pendant polyvinyl alcohol/Fe-Ni alloy nanocomposite. The developed system achieved good analytical performance for phosmet detection and good operability, reproducibility and storage stability. It is worth mentioning that the pre-treatment stage of their method is very simple, involving only liquid-liquid extraction. In another use of composite nanomaterials, a novel biosensor based on elastin-like-polypeptide-organophosphate hydrolase (ELP-OPH), titanium dioxide nanofibers (TiO_2NFs), carboxylic acid-functionalized multi-walled carbon nanotubes (c-MWCNTs) and bovine serum albumin (BSA) was proposed [94]. This biosensor has many advantages, such as rapidity, simplicity, sensitivity and selectivity, for the detection of OPs in aqueous systems. Furthermore, Yu et al. developed a novel biosensor based on the immobilization of AChE on palladium-reduced graphene nanocomposites (rGO@Pd)-modified glassy carbon electrode (GCE) was developed [95]. Their developed system provides a novel and promising tool for the analysis of enzyme inhibitors. Additional deepening the research into this area, introduced a highly stable electrochemical AChE biosensor for detection of OPs. This biosensor was developed by adsorption of AChE on a TiO_2 sol-gel, chitosan (CS) and reduced graphene oxide (rGO)-based multi-layered immobilization matrix (denoted as CS@TiO_2-CS/rGO) [96]. This biosensor displayed many advantages including high reliability, simplicity, and rapidity, and its system provides an excellent platform for immobilization of AChE.

6. Bacteria-Based Eletrochemical OP Biosenors

Bacteria-based electrochemical sensors are established by joining this type of microorganisms to the electrochemical electrode (transducers). If we try to compare bacterias and enzymes we can see that bacterias offer numerous advantages over the enzymes for to be used in biosensor application such as low cost (elimination of the time-consuming and expensive processes of extraction of the intracellular enzymes and their purification), ability to catalyze sequential reactions involving multiple enzymes, resistance to pH and temperature changes (the retention of the enzymes in their natural environment), higher tolerance to toxic substances; enzyme activity recovery in nutrient medium [97].

On the other hand, the bacterial electrochemical sensors are less sensitive and less selective than the enzyme ones, and their response time is relatively long, because of the diffusion constraints imposed by the bacterial cell wall. However, these drawbacks could be overcome, by genetic engineering and by cell permeabilizing respectively, applying various techniques [97].

A small number of bacteria-based electrochemical sensors for OP pesticide determination have been developed so far. It includes genetically engineered *Escherichia coli*, *Pseudomonas putida* or *Moraxella sp.* as a biological recognition element that is surface-expressed OP [98, 99]. The detection principle is identical to the one which is described above when the isolated and purified enzyme was engaged. Recently, microbial sensors based on Clark dissolved oxygen electrode modified with recombinant p-nitrophenol degrading/oxidizing bacteria endowed with OPH activity was reported [100]. The surface-displayed OPH catalyzes the hydrolysis of OP pesticides with nitrophenyl substituent to release products, metabolized by the bacteria while consuming oxygen. The oxygen consumption is measured and correlated to the OP concentration.

CONCLUSION

Through the resistance of pests to the current pesticides, the advance and application of new pesticides are expected. Consumption of pesticides in large-scale and long-term has led to their worldwide flow, even spreading to distant polar regions, polluting water and soil, and eventually causing harm to humans. Overall, the toxicity is not acute, but pesticides can still have effects in low concentration. The endocrine disorderly effect of pesticides is a problem that still requests much more investigation. Detection of pesticide pollution is the primary step to deal with the problem of pesticide pollution in the environment; still, the amount of pesticide pollution around the biosphere is blurred, due to lack of monitoring information. The traditional detection methods are highly effective in aqueous environments, but still suffer some faults that prevent their widescale implementation. Therefore the advance in electrochemically enzyme-based pesticide detection technology is necessary, based on nanomaterials that exhibit unique electrical, chemical and physical properties. Biosensors based on enzyme-functionalized nanostructures expose great potential predictions in environmental monitoring applications.

Although the rapid expansion of biosensors, there are still some complications when applying enzyme-based biosensors in the field measurements. The effects of different OP pesticide compounds on the same enzyme are not well enough understood. It has been described that different OPs have similar inhibitory effects on AChE. Therefore, it is impossible to determine which kinds of OPs affect the activity of enzymes in a particular sample. In reality, the samples are likely to have more than one OP or even other interfering substances. This is a difficult and urgent problem to be solved in the practical detection of OPs by enzyme-based biosensors. In the laboratory, the species and concentrations of pesticides in the samples are all known, but this is not the case on-field. Due to the different conditions, the results in the laboratory cannot be validated using the results obtained from real samples. To test the accuracy of the detection results, it is necessary to compare the results of enzyme-based biosensors with the traditional methods, but this is inefficient. Thus, the detection precision of enzyme-based biosensors has become a major difficulty to their practical application. Pesticide sensors research has opened new horizons with the rapid development of nanotechnology and nanomaterials. With the aid of nanomaterials containing CNTs, AuNPs, AgNPs, ZrO2-NPS and CdS NPS as the immobilized matrix, the remarkable enhancement of electrocatalytic activity for the detection of OPs with very high sensitivity has been achieved. Furthermore, the majority of AChE-incorporated nanomaterial matrices demonstrate excellent stability during storage. Moreover, there are some reports that nM–pM amounts of OPs can be detected on several nanomaterial-modified electrode surfaces, which demonstrates the inherent ability of the nanomaterial matrices to promote OP detection. Currently, a major research theme is the long-term stability of nanomaterial-modified sensors, and more progress in this field can be expected in the near future. However, the defects of these nanomaterials are hindering their further development in the field of enzyme-based and bacteria-based biosensors, and further experiments may be needed to modify these materials for improved performance.

ACKNOWLEDGMENTS

This work was supported by the Ministry of Education, Science and Technological Development of the Republic of Serbia (Grants No. 172015). The author would like to thank Dr. Tamara Lazarević-Pašti for her valuable comments and suggestions during the preparation of this Chapter.

REFERENCES

[1] Singh, B. K. (2009). Organophosphorus-degrading bacteria: ecology and industrial applications. *Natural Reviews Microbiology,* 7: 156-164.

[2] Kwong, T. C. (2002). Organophosphate pesticides: biochemistry and clinical toxicology. *Therapeutic Drug Monit,* 24 (I): 144-149.

[3] Jeyaratnam, J. and Maroni, M. (1994). Organophosphorous Compounds. *Toxicology*, 91: 15-27.

[4] Hill, E. F. (2003). Wildlife Toxicology of organophosphorus and carbamate pesticides, In: *Handbook of ecotoxicoligy*, 2nd edition, Hoffman, D. J.; Rattner, B. A.; Burton Jr., G. A. & Cairns Jr., J., (Eds), Lewis Publishers, CRS Press LLC, Boca Raton, Florida, USA.

[5] Roger, J. and Dagnac, T. (2006). Compounds in Water, Soils, Waste, and Air. In *Chromatographic Analysis of the Environment*, 3rd Edition, Nollet, L. (Ed.), Chromatographic Science Series, CRS Press, Taylor and Francis Group, Boca Raton, Florida, USA.

[6] Zhou Y., Zhang, F., Tang, L., Zhang, J., Zeng, G., Luo L., Liu, Y., Wang, P., Peng, B. and Liu, X. (2017). Simultaneous removal of atrazine and copper using polyacrylic acid-functionalized magnetic ordered mesoporous carbon from water: adsorption mechanism. *Scientific Reports* 7: 1–10.

[7] Lu X., Tao L., Song, D., Li Y. and Gao F. (2018). Bimetallic Pd@Au nanorods based ultrasensitive acetylcholinesterase biosensor for determination of organophosphate pesticides. *Sensors and Actuators B: Chemical* 255: 2575–2581.

[8] Tang, W., Ruan, F., Chen, Q., Chen, S., Shao, X., Gao, J. and Yhang, M. (2016). Independent Prognostic Factors for Acute Organophosphorus Pesticide Poisoning. *Respiratory Care* 61(VII): 965-970.

[9] Sanagi M. M., Salleh, S., Ibrahim, W. A. W., Naim, A. A., Hermawan, D., Miskam, M., Hussain, I. and Aboul-Enein, H.Y. (2013). Molecularly imprinted polymer solid-phase extraction for the analysis of organophosphorus pesticides in fruit samples, *Journal of Food Composition. Analysis* 32 (II): 155-161.

[10] Mir A. F., Morteza, B., Mohammad, R. V. and Mehdi B. (2011). Dispersive liquid–liquid microextraction for the analysis of three organophosphorus pesticides in real samples by high performance liquid chromatography-ultraviolet. *Microchimica Acta* 172: 465-470.

[11] Fukuto T. R. (1990). Mechanism of action of organophosphorus and carbamate insecticides. *Environmental Health Perspectives* 87: 245–254.

[12] Nillos M. G., Gan J. and Schlenk D. (2010). Chirality of organophosphorus pesticides: analysis and toxicity. *Journal of Chromatogrphy. B* 878 (XVII-XVIII): 1277–1284.

[13] Čolović, M. B., Krstić, D.Z., Lazarević-Pašti T.D., Bondžić, A.M. and Vasić, V.M. (2013). Acetylcholinesterase inhibitors: pharmacology and toxicology. *Current Neuropharmacology* 11 (III): 315-335.

[14] Jokanović M. and Kosanović, M. (2010). Neurotoxic effects in patients poisoned with organophosphorus pesticides. *Environmental Toxicology and Pharmacology* 29 (III): 195–201.

[15] Xi, H. -J., Wu, R.-P., Liu, J. -J. and Li, Z. –S. (2015). Role of acetylcholinesterase in lung cancer. *Thoracic Cancer* 6 (IV): 390-398.

[16] Bidari A., Ganjali, M. R., Norouzi, P., Hosseini, M. R. and Assadi Y. (2011). Sample preparation method for the analysis of some organophosphorus pesticides residues in tomato by ultrasound-assisted solvent extraction followed by dispersive liquid–liquid microextraction. *Food Chemistry* 126 (IV): 1840–1844.

[17] Lazarević-Pašti, T., A. Leskovac, and Vasić V. (2015). Myeloperoxidase Inhibitors as potential drugs. *Current Drug Metabolism* 16 (III), 168-190.

[18] Montella, I. R., Schama, R. and Valle, D. (2012). The classification of esterases: an important gene family involved in insecticide resistance - a review. *Memorias do Instituto Oswaldo Cruz* 107 (IV): 437-449.

[19] Marrs, T., Maynard, R. and Sidell, F. (2007). *Chemical warfare agents: toxicology and treatment*. New York: John Wiley & Sons.

[20] Shi, Q., Yuanjie, T., Zhang, Y. and Liu, W. (2018). Rapid detection of organophosphorus pesticide residue on Prussian blue modified dual-channel screen-printed electrodes combing with portable potentiostat. *Chinese Chemical Letters* 29 (VII): 1379-1382.

[21] Fest, C. and Schmidt K. (2002). *The chemistry of organophosphorus pesticides*. Springer Science & Business Media.

[22] Tang, W. and Wu, J. (2014). Amperometric determination of organophosphorus pesticide by silver electrode using an acetylcholinesterase inhibition method. *Analytical Methods* 6: 924-929.

[23] Gupta, R. C. (2005). *Toxicology of organophosphate & carbamate compounds*, 1st ed. Elsevier Academic Press, London.

[24] Corbett, J. R., Wright, K. and Baillie, A. C. (1984). *The biochemical mode of action of pesticides*, 2nd ed., Academic press, London.

[25] Gupta, R. C. (2015) *Handbook of toxicology of chemical warfare agents*. Academic Press, London.

[26] Sanne, B., Mykletun, A., Dahl, A. A., Moen, B.E. and Tell, G.S. (2003). Occupational differences in levels of anxiety and depression: The Hordaland Health Study. *Journal of Occupational and Environmental Medicine* 45 (VI), 628-638.

[27] Compton J. A. (1988). *Military Chemical and Biological Agents*. Telford Press, New Jersey.

[28] Ballantyne, T. and Marrs, T. C. (1992). *Clinical and Experimental Toxicology of Organophosphates and Carbamates*. Butterworth-Heinemann, Oxford, Boston.

[29] Hodges L. C., Bergerson, J. S., Hunter D. S. and Walker C. L. (2000). Estrogenic effects of organochlorine pesticides on uterine leiomyoma cells *in vitro*. *Toxicological Science* 54: 355–364.

[30] Zhao H., Ji, X., Wang, B., Wang, N., Li, X., Ni, R. and Ren, J. (2015). An ultrasensitive acetylcholinesterase biosensor based on reduced graphene oxide-Au nanoparticles-beta-cyclodextrin/Prussian blue-chitosan nanocomposites for organophosphorus pesticides detection. *Biosensors and Bioelectronics* 65: 23–30.

[31] Andersen H. R., Vinggaard, A. M., Rasmussen, T. H., Gjermandsen I. M. and Bonefeld-Jørgensen, E. C. (2002). Effects of currently used pesticides in assays for estrogenicity, androgenicity, and aromatase activity *in vitro*. *Toxicology and Applied Pharmacology* 179 (I): 1–12.

[32] Kojima H., Katsura, E., Takeuchi, S., Niiyama, K. and Kobayashi, K. (2004). Screening for estrogen and androgen receptor activities in 200 pesticides by in vitro reporter gene assays using Chinese hamster ovary cells. *Environmental Health Perspectives* 112 (V): 524–531.

[33] Li J., Li, N., Ma, M., Giesy J. P. and Wang Z. J. (2008). *In vitro* profiling of the endocrine disrupting potency of organochlorine pesticides. *Toxicology Letters* 183 (I-III), 65–71.

[34] Tabb M. M. and Blumberg B. (2006). New modes of action for endocrine-disrupting chemicals. *Molecular Endocrinology* 20: 475-482.

[35] Chapalamadugu S. and Chaudhry G. R. (1992). Microbiological and Biotechnological Aspects of Metabolism Carbamates and Organophosphates. *Critical Reviews in Biotechnology* 12 (V-VI): 357-389.

[36] Salvi, R. M., Lara, D. R., Ghisolfi, E. S., Portela, L. V., Dias, R. D. and Souza D. O. (2003). Neuropsychiatric evaluation in subjects chronically exposed to organophosphate pesticides. *Toxicological Sciences* 72 (II): 267-271.

[37] Yan X., Li, H., Han X. and Su X. (2015). A ratiometric fluorescent quantum dots based biosensor for organophosphorus pesticides detection by inner-filter effect. *Biosensors and. Bioelectronics* 74: 277–283.

[38] Syshchyk O., Skryshevsky, V. A., Soldatkin O. O. and Soldatkin A. P. (2015). Enzyme biosensor systems based on porous silicon photoluminescence for detection of glucose, urea and heavy metals. *Biosensors and Bioelectronics* 66: 89–94.

[39] Lei, Y., Sun, R., Zhang, X., Feng X. and Jiang L. (2016). Oxygen-Rich Enzyme Biosensor Based on Superhydrophobic Electrode. *Advanced Materials* 28 (VII): 1477–1481.

[40] Zhou Y., Tang, L., Zeng, G., Chen, J., Cai, Y., Zhang, Y., Yang, G., Liu, Y., Zhang, C., Tang, W. (2014). Mesoporous carbon nitride based biosensor for highly sensitive and selective analysis of phenol and catechol in compost bioremediation, *Biosensors and Bioelectronics* 61: 519–525.

[41] Zhang W. Y., W. Y., A. M. Asiri, D. L. Liu, D. Du and Y. H. Lin, 2014. Nanomaterial-based biosensors for environmental and biological monitoring of organophosphorus pesticides and nerve agents, *TrAC, Trends in Analytical Chemistry*, 54, 1–10.

[42] Pundir C. S. and Chauhan, N. (2012). Acetylcholinesterase inhibition-based biosensors for pesticide determination: A review. *Analytical Biochemistry* 429 (I): 19-31.

[43] Thévenot D. R., Tóth, K., Durst, R. A. and Wilson, G. S. (1999). Electrochemical Biosensors: recommended refinitions and classification. *Biosensors and. Bioelectronics* 16 (I-II): 121-131.

[44] Liu, G., Wang, J., Barry, R., Petersen, C., Timchalk, C., Gassman P. L. and Lin, Y. (2008). Nanoparticles-based electrochemical immunosensor for the dtection of phosphorylated acetylcholinesterase: an exposure biomarker of organophosphate pesticides and nerve agents. *Chemistry* 14 (XXXII), 9951-9959.

[45] Potkonjak, N. I. (2018). Consideration about a voltammogram as the bifurcation diagram of oscillating electrochemical systems: a case study of the copper|1 M trifluoroacetic acid oscillator. *Reaction Kinetics, Mechanisms and Catalysis* 123: 155-163.

[46] Potkonjak, N. I., Nikolić, Z. M., Anić S. R., and Minić, D. M. (2014). Electrochemical oscillations during copper electrodissolution/ passivation in trifluoroacetic acid induced by current interrupt method. *Corrosion Science* 83: 355-358.

[47] Potkonjak, N. I., Veselinović, D. S., Novaković M. M., Gorjanović, S. Ž., Pezo L. L. and Sužnjević, D. Ž. (2012). Antioxidant activity of propolis extracts from Serbia: A polarographic approach. *Food and Chemical Toxicology* 50 (X): 3614-3618.

[48] Milić, S. Z., Potkonjak, N. I., Gorjanović, S. Ž., Veljović-Jovanović, S. D., Pastor F. T., and Sužnjević, D. Ž. (2011). A Polarographic Study of Chlorogenic Acid and Its Interaction with Some Heavy Metal Ions. *Electroanaysis* 23 (XII): 2935-2940.

[49] Potkonjak, N. I., Potkonjak, T. N., Blagojević, S. N., Dudić B. D. and Randjelović, D. V. (2010). Current oscillations during the anodic dissolution of copper in trifluoroacetic acid. *Corrosion Science* 52: 1618-1624.

[50] Gorjanovic S. Ž., Novaković, M. M., Potkonjak N. I. and Sužnjević, D. Ž. (2010). Antioxidant activity of wines determined by a polarographic assay based on hydrogen peroxide scavenge. *Journal of Agricultural ad Food Chemistry* 58 (VIII): 4626-4631.

[51] Lazarević-Pasti, T. D., Bondžić, A. M., Pašti, I. A., Mentus, S. V. and Vasić, V. M. (2013). Electrochemical oxidation of diazinon in

aqueous solutions via electrogenerated halogens - Diazinon fate and implications for its detection. *Journal of Electroanalytical Chemistry* 692: 40-45.

[52] Gorjanović S. Ž., Novaković, M. M., Potkonjak, N. I., Leskosek-Čukalović I. and Suznjević, D. Ž. (2010). Application of a novel antioxidative assay in beer analysis and brewing process monitoring. *Journal of Agricultural ad Food Chemistry* 58 (II): 744-751.

[53] Lazarević-Pašti, T., Nastasijević, B. and Vasic, V. (2011). Oxidation of chlorpyrifos, azinphos-methyl and phorate by myeloperoxidase. *Pesticide Biochemistry and Physiology* 101 (III): 220-226.

[54] Lazarević-Pašti, T. and Vasić, V. (2010). Oxidation of diazinon for the sensitive detection by cholinesterase-based bioanalytical method. *Journal of Environmental Protection and Ecology* 12 (III): 1168-1173.

[55] Andreescu, S. and Marty, J. L. (2006). Twenty years research in cholinesterase biosensors: From basic research to practical applications. *Biomolecular Engineering* 23 (I): 1-15.

[56] Supraja, P., Tripathy, S., Vanjari, S. R. K., Singh V. and Singh, S. G. (2019). Electrospun tin (IV) oxide nanofiber based electrochemical sensor for ultra-sensitive and selective detection of atrazine in water at trace levels. *Biosensors and Bioelectronics* 141: doi.org/10.1016/j. bios.2019.111441.

[57] Rassaei, L., Marken, F., Sillanpää, M., Amiri, M., Cirtiu C. M. and Sillanpää, M. (2011). Nanoparticles in electrochemical sensors for environmental monitoring. *TrAC Trends in Analytical Chemistry* 30 (XI): 1704-1715.

[58] Rajangam, B., Daniel D. K. and Krastanov, A. I. (2018). Progress in enzyme inhibition based detection of pesticides. *Engineering in Life Science* 18: 4–19.

[59] Andreescu, S., Noguer, T., Magearu V. and Marty J. -L. (2002). Screen-printed electrode based on AChE for the detection of pesticides in presence of organic solvents. *Talanta* 57 (I): 169–176.

[60] Gahlaut, A., Gothwal, A., Chhillar A. K. and Hooda, V. (2012). Electrochemical Biosensors for Determination of Organophosphorus Compounds: Review. *Open Journal of Applied Biosensor* 1: 1-8.

[61] Pundir, C. S. and Chauhan, N. (2012). Acetylcholinesterase inhibition-based biosensors for pesticide determination: A review. *Analytical Biochemistry* 429 (I): 19–31.

[62] Campanella, L., M. Achilli, M. P. Sammartino and T. Mauro, 1991. Butyrylcholine enzyme sensor for determining organophosphorus inhibitors *Bioelectrochemistry and Bioenergetics* 26, 237–249.

[63] Campanella, L., De Luca, S., Sammartino M. P. and Mauro, T. (1999). A new organic phase enzyme electrode for the analysis of organophosphorus pesticides and carbamates. *Analytica Chimica Acta* 385 (I-III): 59–71.

[64] Arduini, F., Forchielli, M., Amine, A., Neagu, D., Cacciotti, I., Nanni, F., Moscone, D.and Palleschi, G. (2015). Screen-printed biosensor modified with carbon black nanoparticles for the determination of paraoxon based on the inhibition of butyrylcholinesterase. *Microchim. Acta* 182: 643–651.

[65] Vidal, J. C., Esteban, S., Gil, J. and Castillo, J. R. (2006). A comparative study of immobilization methods of a tyrosinase enzyme on electrodes and their application to the detection of dichlorvos organophosphorus insecticide. *Talanta* 68 (III): 791-799.

[66] Tanimoto de Albuquerqu, Y. D. and Ferreira, L. F. (2007). Amperometric biosensing of carbamate and organophosphate pesticides utilizing screen-printed tyrosinase-modified electrodes. *Analytica Chimica Acta* 596 (II): 210-221.

[67] Sajjadi, S., Ghourchian, H. and Tavakoli, H. (2009). Choline oxidase as a selective recognition element for determination of paraoxon. *Biosensors and Bioelectronics* 24 (VIII): 2509-2514.

[68] Campanella, L., Lelo, D., Martini, E. and Tomassetti, M. (2007). Organophosphorus and carbamate pesticide analysis using an inhibition tyrosinase organic phase enzyme sensor; comparison by

butyrylcholinesterase + choline oxidase opee and application to natural waters. *Analytica Chimica Acta* 587 (I): 22-32.

[69] Touloupakis, E., Boutopoulos, C., Buonasera, K., Zergioti, I. and Giardi, M. T. (2012). A photosynthetic biosensor with enhanced electron transfer generation realized by laser printing technology. *Analytical and Bioanalytical Chemistry* 402: 3237-3244.

[70] Pabbi, M., Kaur, A., Mittal S. K. and Jindal, R. (2018). A surface expressed alkaline phosphatase biosensor modified with flower shaped ZnO for the detection of chlorpyrifos. *Sensors and Actuators B: Chemistry* 258: 215-227.

[71] Lvnitskii, D. M. and Rishpon J. (1994). A potentiometric biosensor for pesticides based on the thiocholine hexacyanoferrate (III) reaction. *Biosensors and Bioelectronics* 9 (VIII): 569-576.

[72] Ivanov, A. N., Evtugyn, G. A., Gyurcsányi, R. E., Tóth, K. and Budnikov, H. C. (2000). Comparative investigation of electrochemical cholinesterase biosensors for pesticide determination. *Analytica Chimica Acta* 404 (I): 55-65.

[73] Toshihiko, I. and Nobuhiko, I. (1995). Potentiometric butyrylcholine sensor for organophosphate pesticides. *Biosensors and Bioelectronics* 10 (V): 435-441.

[74] Khaled, E., Hassan H. N. A., Mohamed G. G., Ragab, F.A. and Seleim, A. E. A (2010). Disposable potentiometric sensors for monitoring cholinesterase activity. *Talanta* 83 (II): 357–363.

[75] Du D., Wang, L., Shao, Y., Wang, J., Engelhard, M. H. and Lin, Y. H. (2011). Functionalized Graphene Oxide as a Nanocarrier in a Multienzyme Labeling Amplification Strategy for Ultrasensitive Electrochemical Immunoassay of Phosphorylated p53 (S392). *Analytical Chemistry* 83 (III): 746-752.

[76] Zheng M., Jagota, A., Strano, M. S., Santos, A. P., Barone P. and Chou, S. G.et al., (2003). Structure-based carbon nanotube sorting by sequence-dependent DNA assembly. *Science* 302: 1545-1548.

[77] Lin S., Liu C. -C. and Chou, T. -C. (2004). Amperometric acetylcholine sensor catalyzed by nickel anode electrode. *Biosensors and Bioelectronics* 20 (I): 9-14.

[78] Upadhyay S., Rao, G. R., Sharma, M. K., Bhattacharya, B. K., Rao V. K. and Vijayaraghavan, R. (2009). Immobilization of acetylcholinesterase-choline oxidase on a gold–platinum bimetallic nanoparticles modified glassy carbon electrode for the sensitive detection of organophosphate pesticides, carbamates and nerve agents. *Biosensors and. Bioelectronics* 25 (IV): 832-838.

[79] Jha N. and Ramaprabhu, S. (2010). Development of Au nanoparticles dispersed carbon nanotube-based biosensor for the detection of paraoxon. *Nanoscale* 2: 806-810.

[80] Sun X. and Wang, X. (2010). Acetylcholinesterase biosensor based on Prussian blue-modified electrode for detecting organophosphorus pesticides. *Biosensors and Bioelectronics* 25 (XII): 2611 - 2614.

[81] Yu G., Wu, W., Zhao, Q., Wei X., and Lu, Q. (2015). Efficient immobilization of acetylcholinesterase onto amino functionalized carbon nanotubes for the fabrication of high sensitive organophosphorus pesticides biosensors. *Biosensors and Bioelectronics* 68: 288 - 294.

[82] Kesik M., Kanik, F. E., Turan, J., Kolb, M., Timur S., Bahadir, M. and Toppare L. (2014). An acetylcholinesterase biosensor based on a conducting polymer using multiwalled carbon nanotubes for amperometric detection of organophosphorus pesticides. *Sensors and Actuators B: Chemical* 205: 39 - 49.

[83] Daniel M. -C. and Astruc, D. (2004). Gold nanoparticles: assembly, supramolecular chemistry, quantum-size-related properties, and applications toward biology, catalysis, and nanotechnology. *Chemical Reviews* 104 (I): 293 - 346.

[84] Du D., Wang, J., Lu, D., Dohnalkova A. and Lin, Y. (2011). Multiplexed electrochemical immunoassay of phosphorylated proteins based on enzyme-functionalized gold nanorod labels and electric field-driven acceleration. *Analytical Chemistry* 83 (XVII): 6580-6585.

[85] Islam M. A., Xia, Y., Steigerwald, M. L., Yin, M., Liu Z. and O'Brien, S., Levicky, R and Herman, I.P. (2003). Addition, suppression, and

inhibition in the electrophoretic deposition of nanocrystal mixture films for CdSe nanocrystals with γ-Fe2O3 and Au nanocrystals. *Nano Letters* 3 (XI): 1603-1606.

[86] Ion A. C., Ion, I., Culetu, A., Gherase, D., Moldovan C. A., Iosub, R. and Dinescu A. (2010). Acetylcholinesterase voltammetric biosensors based on carbon nanostructure-chitosan composite material for organophosphate pesticides. *Material Science and Engineering: C* 30 (VI): 817-821.

[87] Ren X., Meng, X., Chen, D., Tang F. and Jiao, J. (2005). Using silver nanoparticle to enhance current response of biosensor. *Biosensors and Bioelectronics* 21 (III): 433-437.

[88] Ren C., Song, Y., Li Z. and Zhu, G. (2005). Hydrogen peroxide sensor based on horseradish peroxidase immobilized on a silver nanoparticles/cysteamine/gold electrode. *Analytical and Bioanalytical Chememistry* 381: 1179-1185.

[89] Li Z., Wang, Y., Ni Y. and Kokot, S. (2014). Unmodified silver nanoparticles for rapid analysis of the organophosphorus pesticide, dipterex, often found in different waters. *Sensors and Actuators B: Chemical* 193: 205-211.

[90] Liu S. -Q., Xu, J. -J. and Chen, H. -Y. (2002). ZrO_2 gel-derived DNA-modified electrode and the effect of lanthanide on its electron transfer behavior. *Bioelectrochemistry* 57 (II): 149-154.

[91] Zhou J. -H., Deng, C. -Y., Si S. -H. and Wang, S. -E. (2011). Zirconia electrodeposited on a self-assembled monolayer on a gold electrode for sensitive determination of parathion. *Microchimica Acta* 172: 207-215.

[92] Zhang L., Zhang, A., Du D. and Lin, Y. (2012). Biosensor based on Prussian blue nanocubes/reduced graphene oxide nanocomposite for detection of organophosphorus pesticides. *Nanoscale* 4: 4674-4679.

[93] El-Moghazy A. Y., Soliman, E. A., Ibrahim, H. Z., Noguer, T., Marty J. L. and Istamboulie G. (2016). Ultra-sensitive biosensor based on genetically engineered acetylcholinesterase immobilized in

poly(vinyl alcohol)/Fe–Ni alloy nanocomposite for phosmet detection in olive oil. *Food Chemistry* 203: 73-78.

[94] Bao J., Hou, C., Dong, Q., Ma, X., Chen J., Huo, D., Yand, M., Hussein Abd El Galil, K., Chen, W. and Lei, Y. (2016). ELP-OPH/BSA/TiO_2 nanofibers/c-MWCNTs based biosensor for sensitive and selective determination of p-nitrophenyl substituted organophosphate pesticides in aqueous system. *Biosensors and Bioelectronics* 85: 935-942.

[95] Zang, Y, Xia, Z., Li, Q., Gui, G., Zhao, G.and Lin, L. (2017). Surface Controlled Electrochemical Sensing of Chlorpyrifos in Pinellia Ternate Based on a One Step Synthesis of Palladium-Reduced Graphene Nanocomposites. *Journal of Electrochemical Society* 164 (II): 48-53.

[96] Cui H. -F., Wu, W. -W., Li, M. -M., Song, X., Lv, Y. and Zhang, T. T. (2018). A highly stable acetylcholinesterase biosensor based on chitosan-TiO_2-graphene nanocomposites for detection of organophosphate pesticides. *Biosensors and Bioelectronnics* 99: 223-229.

[97] D'Souza, S. F. (1989). Immobilized cells: techniques and applications. *Indian Journal of Microbiology* 29: 83-117.

[98] Mulchandani, A., Chen, W., Mulchandani, P., Wang J.and Rogers K. R. (2001). Biosensors for direct determination of organophosphate pesticides. *Biosensors and Bioelectronics* 16 (IV-V): 225-230.

[99] Mulchandani, P., Chen W. and Mulchandani, A. (2006). Microbial biosensor for direct determination of nitrophenyl-substituted organophosphate nerve agents using genetically engineered Moraxella sp. *Analytica Chimica Acta* 568 (I-II): 217-221.

[100] Lei, Y., Mulchandani, P., Chen W.and Mulchandani A. (2005). Direct determination of pnitrophenyl substituent organophosphorus nerve agents using a recombinant *Pseudomonas putida* JS444-modified Clark oxygen electrode. *Journal of Agricultural and Food Chemistry* 53 (III), 524-527.

In: Organophosphate Pesticides
Editor: Fabrice Marquis

ISBN: 978-1-53618-307-8

Chapter 3

COMPUTATIONAL MODELLING OF ORGANOPHOSPHOROUS PESTICIDES – DENSITY FUNCTIONAL THEORY CALCULATIONS

Dragana D. Vasić Aničijević*
Department of Physical Chemistry,
Vinča Institute of Nuclear Sciences,
University of Belgrade, Belgrade, Serbia

ABSTRACT

Application of computational methods to the investigation of organophosphorous pesticides (OPs) represents an inevitable step towards the complete understanding of their behavior in any type of environment. Two major directions are modelling of 1) OP interactions with biomolecules, and 2) their adsorption on inorganic materials. *In silico* investigation of OP interactions with biomolecules is important in the systematic studies of toxicity, for example when the influence of particular

* Corresponding Author's Email: draganav@vin.bg.ac.rs.

functional groups on the overall toxicity is evaluated. Moreover, it is useful when designing potential antidotes and neutralizing agents. On the other hand, investigation of organophosphate interactions with various adsorbent surfaces can provide insight into the efficiency of their removal from the contaminated media - water, soil and air - by adsorption onto various, properly designed substrates. Implementation of *in silico* methods also reduces the exposure of the laboratory staff and equipment to the potential hazards from these highly toxic substances.

Density functional theory (DFT) is a powerful tool for determining of the electronic properties of molecular systems. In investigations of organophosphate pesticides, it has a large-scale application in the estimation of reactivity of the functional groups towards substrates of interest (for example through calculation of Fukui indices). Organophosphate molecule properties, such as geometry, dipole moment, charge distribution, ionization potential and electron affinity, obtained by DFT, are used as descriptors – input parameters for further semiempirical modelling of their interactions with biomolecules. Calculation of adsorption properties on the selected substrates – adsorption geometry, binding strength and charge distribution between the molecule and substrate, are used for the design of innovative, and understanding of existing materials for adsorption, degradation and sensing of organophosphate pesticides.

Keywords: organophosphorous pesticides, density functional theory, toxicity, adsorptive removal, environmental protection

INTRODUCTION

Correlation of the macroscopic manifestations of the OP interaction with the (biological or non-biological) environment, and the fundamental properties of this interaction on the atomic level is important because it opens the possibility to predict behavior which is undetectable or difficult to detect by experiment. Moreover, it opens the novel possibilities to modify and design advanced agents for particular purpose (i.e., antidotes or adsorbents for OP removal) which have not been experimentally tested before. Methods of theoretical chemistry enable the investigation of basic interaction properties, without the impact of external factors which are difficult to control. Combined theoretical and experimental studies are often performed, in order to provide complete insight into a particular interaction. In this manner, it is possible to conduct so-called *in silico experiments* which are equal to the classical ones and provide access to the properties of matter which are unreachable otherwise.

On the other hand, the real system which has been observed in the classical experiment is replaced by a particular model. As every model basically comprises approximation, determining the level of approximation, which is satisfactory for certain investigations, becomes the main challenge for the researchers [1]. Contemporary methods of theoretical chemistry based on classical molecular mechanics (MM) and molecular dynamics (MD) provide a good framework for modelling of interaction of small organic molecules (including OPs) with biomacromolecules (i.e., proteins, but also other biomolecules of particular interest, including nucleic acids). Systems with more complex elemental composition which involve metal/non-metal interactions (for example, interaction of OPs with crystalline metal-based adsorbents or metalloproteins) imply a need of more sophisticated approach including quantum mechanics (QM) or *ab initio* and combined (QM/MM) calculations. Density functional theory (DFT) belongs to the group of *ab initio* methods that take *a priori* only fundamental physical constants.

Organophosphate molecules, as well as the substrates they are acting on (i.e., enzymes or adsorbent surfaces) are complex and contain a large number of atoms. To provide a compromise between the accuracy of the model and the computational costs, it is of a primary interest to carefully design the computational model of a particular system, which should be complex enough to depict the key interactions, and simple enough that calculations can be performed in real time (i.e., from a couple of hours to a couple of days).

Building a Computational Model

In order to simplify a complex binary OP-substrate system (i.e., OP+protein or OP+adsorbent surface) it is of primary interest to find and consider a reactive part of the OP molecule. A common approach for the determination of reactive site of organic molecules is to calculate Fukui indices [2] that provide information on which parts of the molecule are most nucleophilic or most electrophilic. Index of electrophilicity f_a^- and index of nucleophilicity f_a^+, are calculated from the equations:

$$f_a^- = e(neutral) - e(cation) \tag{1}$$

$$f_a^+ = e(anion) - e(neutral) \tag{2}$$

Here, *e(neutral)* is the partial charge of an atom in the neutral molecule, *e(cation)* is the partial charge of the atom when an electron is subtracted from the molecule, and *e(anion)* is the partial charge of an atom when an extra electron is added to the considered molecule. DFT calculations in essence deal with electron density distribution [3] (which is not the case for the above mentioned classical methods) and are an appropriate tool to provide input for calculation of partial atomic charges. A significant number of methods are developed to assign a DFT-obtained charge in the molecule to a particular atom (for example Lowdin partial charges [4], Mulliken population analysis [5] and Bader charge analysis [6]). Many quantum chemical and DFT computational packages have implemented some of these methods, so the charge distribution among the atoms in a molecule is easily available as a part of postprocessing results of a self-consistent calculation [7] for an isolated OP molecule.

An interesting example is recently provided by Lazarević-Pašti and Anićijević [8, 9]. The adsorption of a dimethoate molecule on a graphene surface was studied by a combined experimental and DFT approach. As graphene is considered an electrophilic material, nucleophilicity of dimethoate was analyzed through the calculation of Fukui indices from the difference of Lowdin charges. Results are shown in Table 1:

Table 1. Fukui indices of the most important atoms of a dimethoate molecule. Referent atoms are indicated on the sketch of structural formula by a red circle. For very similar atoms Fukui indices are shown simultaneously, as an interval

atom name	atom position	fa^-	fa^+
O	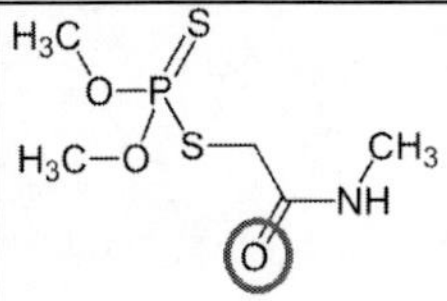	0.10	0.06
C	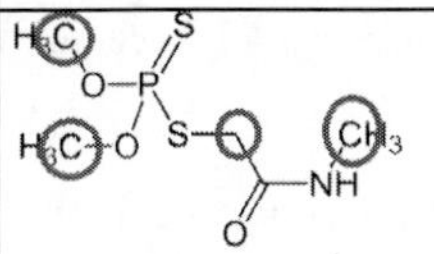	>0.02	>0.01
N	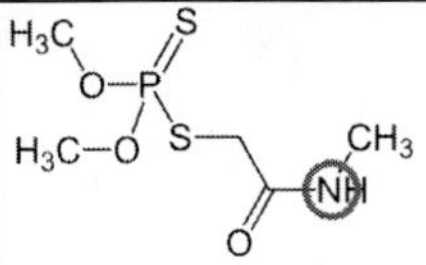	0.04	0.01
S	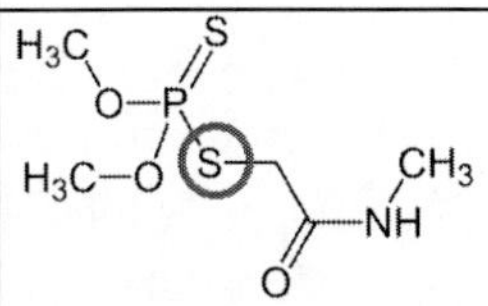	0.14	0.15
P	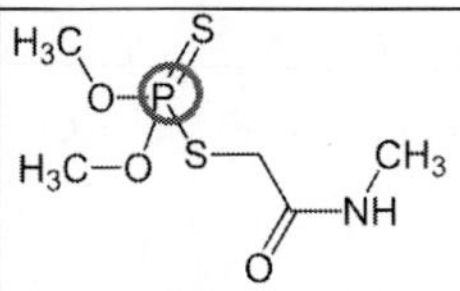	0.05	0.15
S	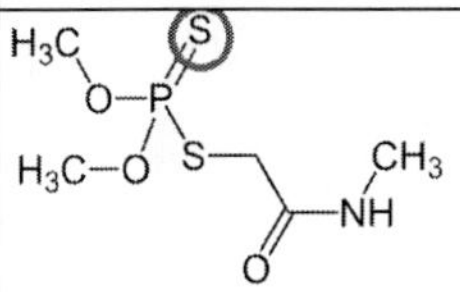	0.30	0.23
O	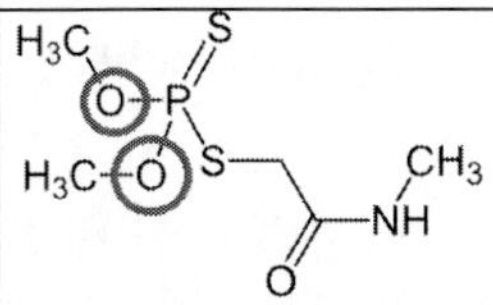	0.05 - 0.06	0.06 - 0.07

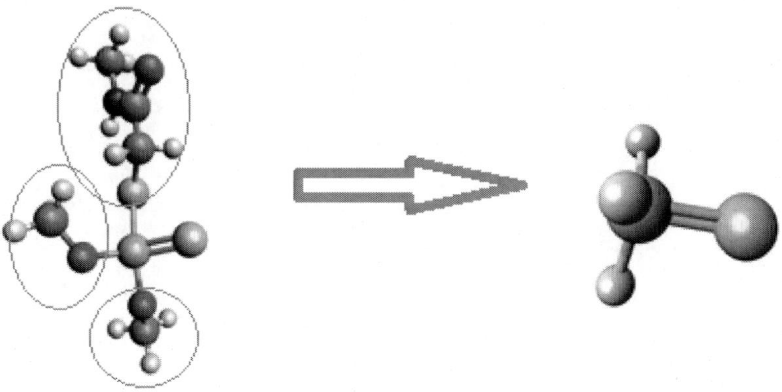

Figure 1. Abstraction of a dimethoate molecule into a model molecule H_3PS (reconstructed from the data given in [8]).

The strongest nucleophile is the sulphur atom which is bound to a central phosphorous atom by a double bond. According to this, it was proposed that the sulphur atom is the one that will be most involved in the nucleophilic attack to graphene. In the subsequently considered model molecule (Figure 1), central phosphorous atom with sulphur was kept, while all other substituents on the phosphorous atom were replaced by hydrogens. The adsorption of model molecule is further considered on different graphene surface models, in order to estimate binding ability of the OP molecule on different graphene surfaces.

This manner of simplification of adsorbate is practical when the chemically complex supports such as carbonaceous materials are considered, to evaluate the impact of various features of the support (i.e., defects, edges, surface groups, oxides…) on the reactivity of certain type of molecules. Aničijević has also proposed 4 different theoretical models of graphene surface – pristine graphene, graphene with a monovacancy defect, graphene with an epoxide defect and graphene with an oxy-defect, to examine their interaction with the designed model molecule. Each proposed model represents a feature that is present in real graphene and graphene-oxide samples, including defects and oxygen functional groups. Model molecule, H_3PS, was then adsorbed on a model surface using periodic DFT-GGA calculations and plane-wave basis set in a 4x4 graphene supercell. Obtained results show that only monovacancy defect graphene exhibits affinity towards the adsorption, while other investigated surfaces are unreactive towards H_3PS model molecule. Adsorption energies of HPS are given in Table 2.

Obtained results are in good agreement with other DFT studies that deal with the reactivity of graphene [10, 11].

Table 2. Adsorption energies of the model molecule H_3PS on various graphene surfaces (adapted from [8])

Surface model		H_3PS adsorption energy/eV
Pristine graphene	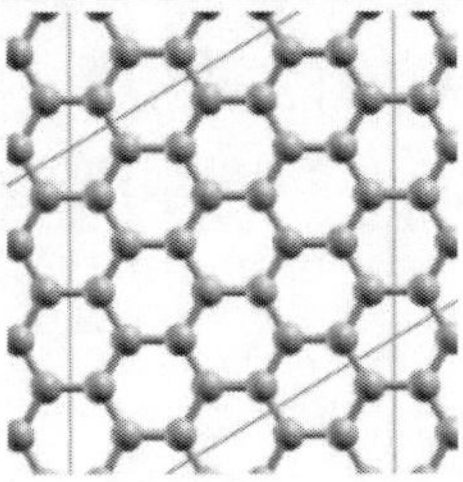	-0.01
Graphene with monovacancy defect	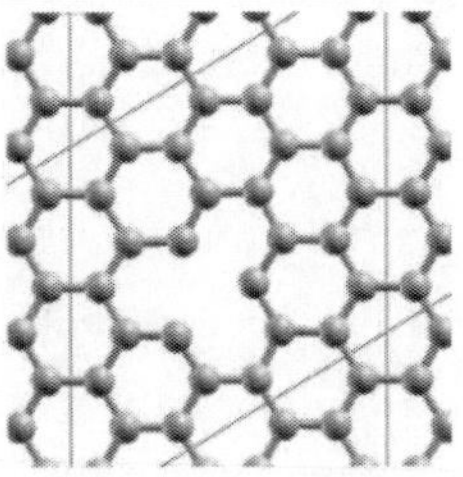	-4.04
Graphene with oxy-defect	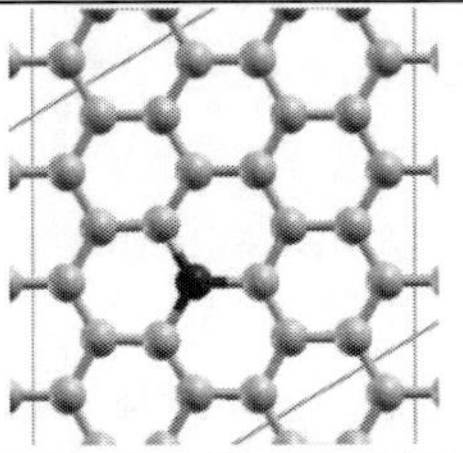	-0.04
Epoxide graphene	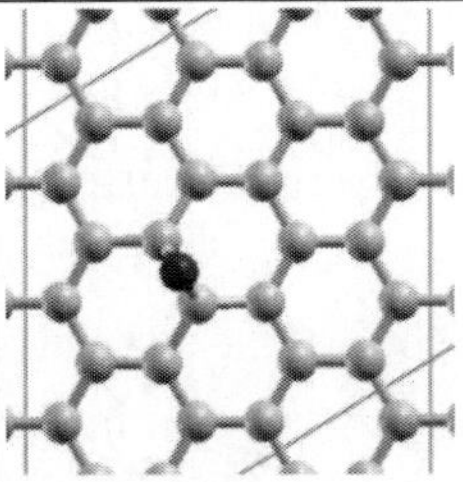	+0.02

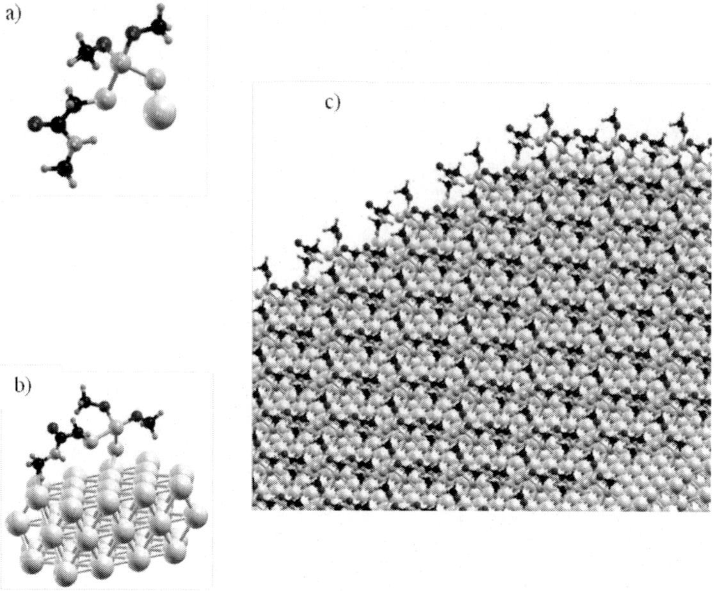

Figure 2. Comparison of the dimethoate adsorption on three different models of metallic Ag as a sample surface (left: 1-atom model, middle: isolated 48-atom cluster and right: Ag(111) surface slab).

When organic molecules (including OPs) are adsorbed non-carbonaceous crystalline and metallic supports, it is possible to introduce some different surface models. So-called 1-atom surface model in case of metallic surfaces is appropriate for determining the reactive part of an organic molecule towards a particular metal [12-15]. The 1-atom surface model is simple and reliable, and can also be used as a part of preliminary calculations in studies of "organic molecule – metallic surface" interaction. However, when the adsorbent surface is more complex (for example, adsorption of the organic molecule on a carbonaceous material, carbide or oxide surface is studied) the 1-atom model is in general too simple to be used and then some additional features should be included into the surface model, making the direct comparison of reactivity different parts of the molecule more difficult. Replacing adsorbate surface by a representative cluster of atom is the next possibility to build the model of adsorbent-adsorbate system. This is where the boundary conditions become particularly important, and the results of calculation are extremely dependent on the size of a cluster and the position of adsorbate [13, 16].

Finally, the slab model is probably the most well-known and most exploited when adsorption is studied. Although it is simpler than the cluster model in view of boundary conditions, and enables control of surface concentration of the investigated molecule, it has some serious disadvantages when the charged molecules are adsorbed [16, 17]. 1-atom, cluster and slab surface models, with the dimethoate molecule as an example adsorbate, are given in Figure 2.

DENSITY FUNCTIONAL THEORY AS A TOOL FOR PREDICTION OF ORGANOPHOSPHATE TOXICITY

The toxicity of organophosphates essentially originates, to the highest degree, from its interaction with the active site of the enzymes – cholinesterases. Theoretical investigations of organophosphate toxicity are closely correlated to the OP – protein interactions. As the size of cholinesterase molecules (approximately 4000 atoms) currently overcomes the scope of the most ambitious DFT investigated systems (few hundreds of atoms) by at least an order of magnitude, it is impossible to fully describe this interaction using only DFT calculations. The size of the human AChE molecule compared to the dimethoate is compared for illustration, in Figure 3. For a quick orientation, one should bear in mind that the amount of computational time required for a DFT calculation scales approximately with N^3, where N is the number of atoms in the system [1, 18]. So, the exact DFT calculations involving cholinesterases would require about 500 - 1000 times more time than the most ambitious DFT calculations of today. One promising way to overcome this problem is based on QM/MM calculations, where the substrate and the active site are treated by quantum mechanical, *ab initio* (including DFT) methods, while the rest is threated by less demanding methods of molecular mechanics. The most complex issues in QM/MM approach remain the possible artefacts that are formed on QM/MM boundary, and still, high computational costs of the calculations. In the last decade, numerous research efforts are performed in the field of developing novel, semiempirical methods, that are particularly intended for the computational investigations of enzyme/substrate interaction [19]. Although it overcomes the scope of this chapter, developing and getting with the additional novel semiempirical methods, seems to be inevitable for straightforward theoretical investigations of enzyme/substrate interface.

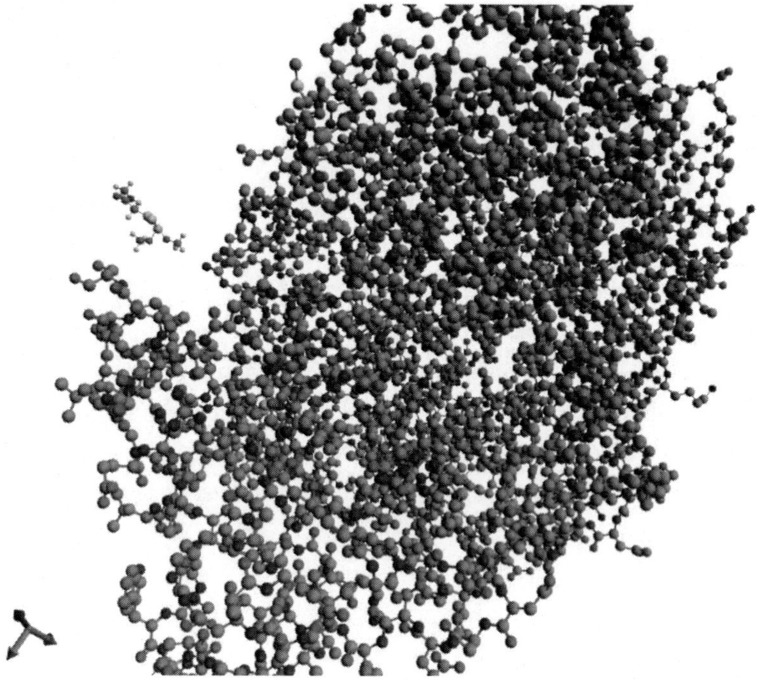

Figure 3. Dimethoate molecule in comparison with human AChE molecule.

However, there are some interesting examples where DFT is employed as an auxiliary tool, and the complement to the experimental observations. Vlahović et al. [20] used density functional method with continuum solvation model for the determination of lipophilicity through the calculation of the partition coefficient between octanol and water, for 22 most frequently used organophosphate type pesticides. Partition coefficient, was calculated from the Gibbs free solvation energies obtained by the use of three different functionals (local M06L, general gradient PBE and hybrid M062X), according to the formula:

$$\log K_{OW} = \frac{\Delta G_{solv(water)} - \Delta G_{solv(octanol)}}{2.303RT} \quad (3)$$

A very good agreement between theoretically determined and experimentally determined partition coefficients was achieved. Obtained results were correlated with median lethal doses (LD50) and median lethal concentrations (LC50) of the investigated pesticides. It was shown that organophosphates with extreme lipophilicity (or hydrophilicity) are less toxic than these with equal distribution between two solvents (medium lipophilicity). As admitted by the authors, this work highlights DFT as an important tool for the lipophilicity calculation of existing pesticides, and this method that can be reliable for the predicting of the lipophilicity of pesticides that have not been synthesized yet.

Mendonca and Snurr [21] calculated the uncatalyzed alkaline hydrolysis mechanism for 117 potential organophosphate simulant molecules. Functional groups and leaving groups were systematically varied to observe the effects on overall reaction energetics. For geometry optimization, B3LYP functional [22] and 6–311^{++}G** basis set were used. Gibbs free energies of transition states ΔG_{TS} for stepwise and concerted S_N2 substitution of various leaving groups by OH- group were calculated. Obtained results were used to identify potential organophosphate simulants of dangerous neurotoxins, having the identical energy barriers for hydrolysis but being less toxic in common experimental conditions. In addition, DFT calculated descriptors (P atom charge, dipole moment, P=O bond order, electron affinity, ionization potential, pKa) were used to build QSAR models [23] to estimate toxicity.

DFT is frequently employed in the studies of biochemical systems as a tool to calculate vibrational spectra of organic molecules which is difficult to obtain experimentally, or for comparison with experimental data, as in case of monocrotophos. In [24] the molecular structural parameters and vibrational frequencies of the fundamental modes of monocrotophos have been obtained using Density functional theory (DFT), using B3LYP functional with 6-311^{++}G(d, p) basis sets and the detailed vibrational analysis of FT-IR and FT-Raman spectral bands have been carried out using potential energy distribution (PED). In geometry optimization, phosphate structure has been found to be distorted, forming three single P-O bonds, and the fourth bond is P=O with a partial double bond character. The observed disruption of the structure was used to discuss shifting of some vibrational modes (asymmetric bending, asymmetric stretching and symmetric stretching) in the vibrational spectrum. The obtained DFT results were further correlated with the results of molecular docking of a monocrotophos molecule in bovine serum albumin (a globular protein) and DNA. Binding was observed in both cases, and it was discussed in light of monocrotophos molecular and electronic characteristics, obtained from DFT calculations.

In [25] DFT was used to compute reaction enthalpies, free energies, and activation barriers *in vacuo*, for the nucleophilic substitution of fluorophosphates, by a wide variety of R_1OH nucleophiles. B3LYP hybrid functional was used, and frequency analysis was performed to confirm the local minima of the potential energy surface. It was found that the substitution via a proton transfer from the nucleophile to the fluorine atom through the phosphinyl oxygen atom (Wright-type mechanism), has a lower activation barrier in the gas-phase than for the corresponding mechanism that operates via a proton transfer from the nucleophile directly to the fluorine atom. Of the nucleophilic agents investigated, peroxybenzoic acid and o-iodosobenzoic acid had the lowest classical activation barrier for the Wright-type mechanism. As stated by the authors, increased understanding of the thermodynamics/kinetics involved in the reactions of OP compounds with potential destructants could lead to the design of superior pesticides with greater safety margins, or prophylactics for bodily exposure to the most toxic organophosphates.

Finally, there is an example of a complex QM/MM study in the field of bioscavengers. Wymore et al. [26] investigated the mechanism of hydrolysis of organophosphate S-sarin catalyzed by the enzyme diisopropyl fluorophosphatase (DFPase), which was previously found to catalyze hydrolysis of organophosphates, with the aim to apply it in prophylactics of organophosphate poisoning. They performed quantum mechanical/molecular mechanical (QM/MM) simulations with density functional theory to investigate and compare the mechanisms of DFP and (S)-sarin hydrolysis by DFPase. The QM region, consisting of 82 atoms for DFP and 75 atoms for S-sarin, included the DFP or (S)-sarin substrate, the side chains of Glu21, Asn120, Asn175, Asp229, Ser271, and Asn272, the catalytic Ca^{2+}, and two crystallographic water molecules coordinated to Ca^{2+}.

Initial geometries for simulations of the hydrolysis of the DFP, in which an activated water molecule attacks Cγ of Asp229, were obtained from DFT/MM potential energy scans. Gas-phase DFT geometry optimizations were performed for acetate–DFP and acetate–(S)-sarin complexes, representative of pentavalent Asp229–substrate complexes, using multiple hybrid functionals and the 6-31^{+}G(d) basis set. Vibrational frequency analysis confirmed that the resulting structures were true energy minima. Umbrella sampling was used for sampling of the preliminary potential energy surface. Simulations were divided into 48 windows and each was equilibrated in PM6/MM semiempirical potential prior to equilibration in DFT/MM potential for 3 ps. Each window was then sampled for 20 ps in an NVT ensemble, and the reaction coordinate value was recorded at every time step. Obtained reaction profiles were represented as a function of energy on the reaction coordinate, which was defined as the mass-weighted distance difference between Oδ(Asp229)–P and P–F. In addition, gas phase potential energy scans were performed by varying the acetate–O–P distance for sarin and DFP to provide the deeper insight into the underlying reactivity of Asp229. According to the findings, hydrolysis of sarin by DFPase requires 28 kJ/mol, which is approximately twice larger activation energy for hydrolysis compared to DFP. The large differences in shape and energy of potential curves were explained mainly by the difference in the ability of the two substrates to accommodate the additional negative charge from the incoming Asp229 nucleophile. Thus, it was concluded that the findings from the QM/MM simulations can be traced to differences in the electrophilicity and reactivity of the phosphorus centers in DFP and (S)-sarin.

The above studies represent illustrative examples the role of DFT in modeling of organophosphate toxicity, which is also summarized in Figure 4:

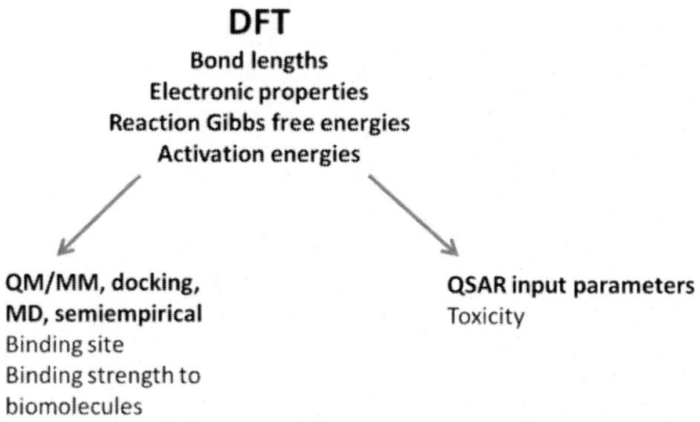

Figure 4. A summary of application of DFT results as input for other investigation methods.

ADSORPTION OF ORGANOPHOSPHATES – STRATEGIES FOR THE REMOVAL FROM THE ENVIRONMENT

The fast and reliable detection of organophosphates and their effective removal from the environment imply a need for the design and development of advanced materials for these particular purposes. From the point of view of theoretical investigations, it is important to know the adsorption properties of organophosphates on the materials intended for sensing or removal. Some representative examples of DFT investigation of organophosphate adsorption, are given in the following text.

In the recent paper of Lazarević-Pašti et al. [9] DFT calculated *in vacuo* charge distribution of organophosphate pesticides chlorpiryphos (CPF) and dimethoate (DMT) was correlated with experimentally observed acetylcholinesterase (AChE) inhibition curves and the adsorption properties on graphene.

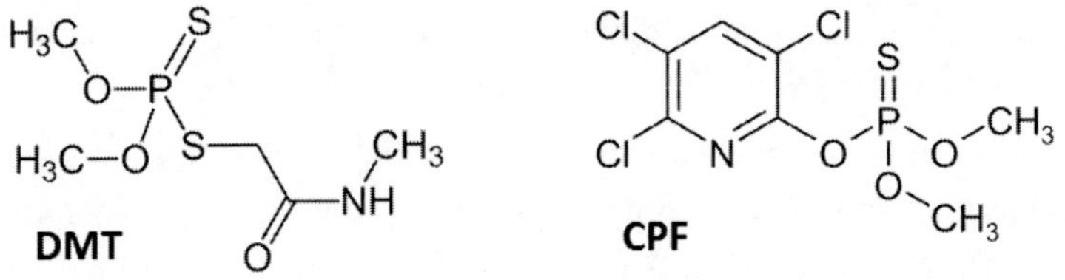

Figure 5. Structural formulas of dimethoate (DMT) and chlorpyriphos (CPF).

AChE inhibition curves of DMT and CPF showed that the (neuro)toxicity of CPF is much higher than that of DMT (what is in agreement with the higher CPF solubility in lipids due to the presence of an aromatic ring). The sulfur atom, forming a double bond with P in both DMT and CPF, bears a negative charge (−0.54 e in both DMT and CPF), while phosphorous is positively charged (~+1.5 e, Figure 1), being susceptible to nucleophilic attacks.

DMT is an aliphatic compound, while CPF has an aromatic ring with three Cl substituents which bear a small negative charge. Structural formulas of CPF and DMT are compared in Figure 5, for clarity. Principal structural differences that determine toxicity, also determine the differences in adsorption properties on graphene between these two compounds, as it is further shown in the paper.

It was proposed by the authors that there are following main types of interaction between the functional groups of investigated organophosphates and the different features on graphene materials: 1) methyl and ethyl groups (in DMT) could interact with graphene basal plane (via weak dispersion interactions with the π electron system) or with polar surface groups (via dipole-dipole interactions and/or -S-H----O hydrogen bonds); 2) the aromatic ring of CPF is expected to interact with the π electron system of the graphene basal plane, while negatively charged chlorine substituents can interact with positively charged groups on the adsorbate surface. As a result, it was found that aliphatic DMT does not adsorb on graphene nanoplatelets (GNP), due to the noble character of conserved π electronic system towards the aliphatic groups. Adsorption of DMT is more effective on the graphene materials with more oxygen defects – industrial graphene (IG) or graphene-oxide (GO). On the other hand, CPF can be effectively adsorbed on all investigated types of graphene materials, as it contains an aromatic π system and polar -Cl substituents.

This work represents an illustrative example of a comprehensive study where many aspects of organophosphate behavior are consolidated – both toxicity and removal possibilities are discussed from the point of view of intrinsic molecular properties of two different organophosphate pesticides. On the other hand, the light is shed on the rather complex nature of interactions that arise between the different features of organophosphate and these of adsorbates. Although one might expect that there is some unique functional group that interacts with adsorbate surfaces (as it does with enzymes!), this is obviously not the case. Some of the observed interactions (i.e., dispersion interactions) are not well described by DFT, and should be additionally included through semiempirical corrections (in this work, Grimme-D3 dispersion correction [27] was applied) or through particular functionals with included dispersion interactions.

A year later, Yardav et al. [28] have performed another combined experimental and DFT study of CPF and malathion adsorption on graphene-oxide. DFT investigation of interaction was employed to better understand the mechanism of OP removal by the adsorption onto GO. A hybrid B3LYP functional with 6-31G (d) basis set was used. The authors performed calculations with different structural models for GO: hydroxyl and epoxy graphene, also pointing to the non-homogeneity of its surface. In order to eliminate boundary effects, GO edges have been with hydrogen atoms. In addition, integral equation formalism polarized continuum model (IFEPCM) [29] was used to represent the solvent effects – water was used as a solvent in all calculations.

Optimized geometries of CPF and malathion adsorbed on proposed GO models imply that the binding occurs through the formation of a hydrogen bond between the hydroxyl or epoxy oxygen on the GO surface, and alkyl hydrogen of the organophosphate, or even ether oxygen (in case of malathion). On the other hand, one might expect that some other (i.e., more nucleophilic) parts of the molecule will be more reactive towards GO. The latter result implies that the binding geometry is not necessarily determined by the expected reactivity of the particular groups, but also by the steric properties of the OP molecule. It also opens an important question – the question of the complexity of geometry optimization. Due to the size of the OPs as adsorbates, which contain tens of atoms and tens of bonds, there are so many degrees of freedom (and local energy minima), that it is impossible to confirm that the DFT-optimized geometry really represents the absolute energy minimum. On the other hand, use of semiempirical methods for calculation of input geometries seems to be susceptible to artefacts, and, to my knowledge, there is no study dealing with semiemprically pre-optimized adsorption geometries further subjected to DFT calculations.

The authors particularly overcame this problem: they considered different configurations of pesticides with respect to adsorbents by placing pesticide at varied distance over graphene containing hydroxyl group and graphene containing epoxide group, and calculated adsorption energies. Calculated adsorption energies are in range 0.04-0.13 eV, and it was found that binding of CPF and malathion to hydroxyl group is stronger than to epoxy group.

Finally, there is an example of a pure theoretical DFT study, with the aim to investigate the possibility of application of Fe-doped boron nitride in adsorption of organophosphates. Farmazandeh and Rezainejad [30] used DFT on the GGA-PBE level, to investigate the adsorption of diazinon, hinosan, chlorpyrifos, and parathion organophosphorus pesticides on the surface of $B_{36}N_{36}$ nanocage and its Fe -doped derivatives. Van der Waals interaction was calculated using Tkatchenko and Scheffler method. Hydration energies were calculated using COSMO [31] model. Fukui indices [2] were used as the descriptor for the potential of overlapping of organophosphate orbitals with the adsorbent surface orbitals. According to calculated Fukui indices (f a^+, f a^-, and f $_0$) for atoms of OPs, it is found that the largest fa^-(up to 0.466) are obtained for sulfur atoms, so investigated molecules are mainly nucleophile, while the highest potential for electrophilic attack (f a^+ up to 0.176) is exhibited by aryl carbons and chlorine atoms. Fukui indices and atomic partial charges were considered to find the best sites of OPs for adsorbing on $B_{36}N_{36}$ and FeBN nanocages. Hence, it has been shown that thiophosphate groups in all OPs, aromatic ring in diazinon, and nitro group in parathion are most likely to bind to the mentioned nanocages. Observed adsorption energies are different for different pesticides, and vary in range 1.70–2.68 eV, implying that a real chemisorption occurs.

Conclusion and Perspectives

Density functional theory is a powerful technique for calculation of the electronic properties of molecules. It is most efficient for determination of intrinsic properties of organophosphate molecules, which are further applied as descriptors in semiempirical models of enzyme-substrate interaction or quantitative structure – activity relationship (QSAR) models. The most prominent challenge at the moment in the field of enzyme/substrate interaction studies are QM(DFT)/MM calculations, which enable to view an integral picture of the enzyme/substrate system, but require high-performance computational resources (or, at least, high level of skills of the experimenter), and bear some risk of QM/MM boundary artefacts, due to the enormous complexity of the system on the atomic level. Nevertheless, there are great examples in this field, shedding light on the future possibilities to use pure *in silico* experiments, for example, in designing of bioscavengers.

DFT is also an important technique to study interaction of organophosphates with various adsorbing surfaces, with the aim to employ studied materials as adsorbents for removal or in sensing. Particular attention in this chapter was paid to the carbonaceous support and modelling of their most prominent features – defects and oxygen groups. Contrary to the interactions with enzymes where the active site is specific for the phosphate group, active parts of the OP molecule that will interact with the surface are not always easy to determine. Discussed literature examples show that there are significant differences that can be observed in adsorption properties, even when minor features of adsorbate (or adsorbent surface) are modified. Frequent method to find the reactive group is the calculation of the Fukui indices from charge distribution. However, this method does not take into account steric properties of, relatively large, organophosphate molecules. Determination of preferential adsorption geometries seems to be the main challenge in the field of OP adsorption. At the moment, except calculation of the Fukui indices, there are no reliable algorithms to predict the preferential adsorption geometry of large organic molecules (including OPs). There is an increasing number of DFT studies, including pure computational and combined experimental and theoretical ones, that deal with the adsorption of OP pesticides on various supports, implying to the important role of *in silico* experiments in design of the materials for their adsorptive removal.

ACKNOWLEDGMENTS

The author would like to thank the Ministry of Education and Science of the Republic of Serbia (Project No. 172045) for their financial support.

REFERENCES

[1] Vasić Aničijević, Dragana.2015. "*Theoretical analysis of tungsten carbide properties as an electrocatalyst support for hydrogen electrode reactions.*" PhD diss., University of Belgrade.

[2] Ayers, P. W.; Yang, W.; Bartolotti, L. J..2010. "Fukui Function." In *Chemical Reactivity Theory: A DFT View*, edited by Chatteraj, P. K. Boca Raton: CRC Press, 269-281.

[3] Schatz, Jürgen.2004.: "Ab Initio Calculations on Supramolecular Systems." In *Encyclopedia of Supramolecular Chemistry*, edited by Jerry L. Atwood, Boca Raton: CRC Press, 1-8.

[4] Löwdin P. O., 1970. "On the non-orthogonality problem" *Advances in Quantum Chemistry*, 5:185–199.

[5] Mulliken, R. S. 1955. "Electronic Population Analysis on LCAO–MO Molecular Wave Functions. I" *J. Chem. Phys*. 23, 1833 https://doi.org/10.1063/1.1740588.

[6] Bader, R. F. W. 1994. *Atoms in Molecules: A Quantum Theory*; Oxford University Press: Oxford, UK; pp. 1–264.

[7] Folland, Nathan O. 1967. 'Self-Consistent Calculations of the Energy Band Structure of Mg2Si, *hys. Rev.* 158, 764.

[8] Anićijević, Vladan J. 2016. "*Uklanjanje organofosfata iz životne sredine putem adsorpcije na grafen i grafen-oksid* [*Organophosphate removal from the environment by the adsorption on graphene and graphene oxide*]" (MSc thesis, University of Begrade, Faculty of Physical Chemistry).

[9] Lazarević-Pašti, Tamara., Anićijević, Vladan., Baljozović, Miloš., Vasić Anićijević Dragana., Gutić, Sanjin., Vasić, Vesna., Skorodumova, Natalia V., and Pašti, Igor A. 2018., "The impact of the structure of graphene-based materials on the removal of organophosphorus pesticides from water" *Environ. Sci.: Nano*, 5: 1482-1494 https://doi.org/10.1039/C8EN00171E.

[10] Vasic Anicijevic, Dragana., Perovic, Ivana., Maslovara, Sladjana., Brkovic, Snezana., Zugic, Dragana., Lausevic, Zoran., Marceta Kaninski, Milica., 2016. "Ab initio study of graphene interaction with O2, O and O-." *Macedonian Journal of Chemistry and Chemical Engineering*, 35: 271-274 doi: 10.20450/mjcce.2016. 1038.

[11] Dobrota, Ana S. Pašti, Igor A., Mentus, Slavko V., and Skorodumova, Natalia V. 2016. "A general view on the reactivity of the oxygen-functionalized graphene basal plane," *Phys. Chem. Chem. Phys.*, 18: 6580-6586. https://doi.org/10.1039/C5CP07612A

[12] Laban, B. Zekovic, I., Vasic Anicijevic, D., Markovic, M., Vodnik, V., Luce, M., Cricenti, A., Dramicanin, M., Vasic, V., 2016. "Mechanism of 3,3′-Disulfopropyl-5,5′-Dichlorothiacyanine Anion Interaction With Citrate-Capped Silver Nanoparticles: Adsorption and J-Aggregation," *J. Phys. Chem. C* 120, 32: 18066-18074 https://doi.org/ 10.1021/acs.jpcc.6b05124.

[13] Ralević Uroš, Isić Goran, Vasić Anicijević Dragana, Laban Bojana, Bogdanović Una, Lazović Vladimir M., Vodnik Vesna, Gajić Radoš. 2018. "*Nanospectroscopy of thiacyanine dye molecules adsorbed on*

silver nanoparticle clusters," 434: 540-548 https://doi.org/10.1016/ j.apsusc. 2017.10.148.

[14] Laban, B. Ralević U., Petrović S., Leskovac A., Vasić-Anićijević D., Marković M., Vasić V. 2020. Green synthesis and characterization of nontoxic L-methionine capped silver and gold nanoparticles, *Journal of Inorganic Biochemistry* 204: 110958 https://doi.org/10.1016/j. jinorgbio.2019.110958.

[15] Binaymotlagh Roya, Farrokhpour Hossein, Hadadzadeh Hassan, Mirahmadi-Zare Seyede Zohreh, Amirghofran Zahra. 2017. "Combined Experimental and Computational Study of the In Situ Adsorption of Piroxicam Anions on the Laser-Generated Gold Nanoparticles" *J. Phys. Chem.* C 121, 15: 8589-8600 https://doi.org/ 10.1021/acs.jpcc.6b12962.

[16] Vasić Anićijević, Dragana D. 2019. "Citrate adsorption on Ag surfaces – establishing a DFT model," *Eighteenth Young Researchers' Conference - Materials Science and Engineering*, December 4-6, Belgrade, Serbia, Book of Abstracts p. 24.

[17] Bal, Kristof M. and Neyts, Erik C. 2018. "Modelling molecular adsorption on charged or polarized surfaces: a critical flaw in common approaches" *Phys. Chem. Chem. Phys.,* 20: 8456-8459 https://doi.org/ 10.1039/C7CP08209F.

[18] Rauk, Arvi. 2001. "*Orbital interaction theory of organic chemistry*" New York:Wiley, 246.

[19] Purg, Miha. 2018. "*Computational modeling of the Mechanisms and Selectivity of Organophosphate Hydrolases*" PhD diss., Uppsala University.

[20] Vlahović Filip Ž., Ivanović Saša, Zlatar Matija, Gruden Maja. 2017. Density functional theory calculation of lipophilicity for organophosphate type pesticides, *Journal of the Serbian Chemical Society*, 82, 12: 1369-1378, 10.2298/JSC170725104V.

[21] Mendonca, Matthew L., Snurr, Randall Q. 2019. "Screening for Improved Nerve Agent Simulants and Insights into Organophosphate Hydrolysis Reactions from DFT and QSAR Modeling" *Chemistry A*

European Journal, 25,39: 9217-9229 https://doi.org/10.1002/chem.201900655.

[22] Lee, Chengteh, Yang, Weitao and Parr, Robert G. 1988. “Development of the Colle-Salvetti correlation-energy formula into a functional of the electron density” *Phys. Rev. B* 37: 785–789. https://doi.org/10.1103/PhysRevB.37.785.

[23] Gini, G. 2016. “QSAR methods,” *Methods Mol Biol*. 1425:1-20. doi: 10.1007/978-1-4939-3609-0_1.

[24] Nimmi, D. E., Chandhini Sam S. P., Praveen, S. G. and Binoy J. 2018. “DFT based Vibrational Spectroscopic Investigations and Biological Activity of Toxic Material Monocrotophos” *AIP Conference Proceedings* 1953: 080029 https://doi.org/10.1063/1. 5032835.

[25] Bock, Charles W. Larkin Joseph D., Hirsch, Stephen S., Wright, J.B. 2009. “Nucleophilic destruction of organophosphate toxins: A computational investigation,” *Journal of Molecular Structure THEOCHEM* 915(1-3): 11-19 10.1016/j.theochem.2009.07.048.

[26] Wymore, Troy, Field, Martin J., Langan, Paul,. Smith, Jeremy C, and Parks Jerry M. 2014. “Hydrolysis of DFP and the Nerve Agent (S)-Sarin by DFPase Proceeds along Two Different Reaction Pathways: Implications for Engineering Bioscavengers” *J. Phys. Chem.* B 118, 17: 4479-4489 https://doi.org/10.1021/jp410422c.

[27] Grimme, Stefan. 2006. “Semiempirical GGA-type density functional constructed with a long-range dispersion correction” *J Comput Chem* 27: 1787–1799 https://doi.org/10.1002/jcc.20495.

[28] Yardav, S., Goel, N., Kumar, V. and Singhal., S. 2019. “Graphene Oxide as Proficient Adsorbent for the Removal of Harmful Pesticides: Comprehensive Experimental Cum DFT Investigations,” *Analytical Chemistry Letters*, 9, 3: 291 -310 https://doi.org/10.1080/22297928.2019.1629999.

[29] Travlou, N.A., Kyzas, G.Z., Lazaridis, N.K. and Deliyanni, E.A. 2013. “Functionalization of graphite oxide with magnetic chitosan for the preparation of a nanocomposite dye adsorbent.” *Langmuir*. 29: 1657–1668. doi: 10.1021/la304696y

[30] Farmanzadeh, Davood., Rezainejad, Hamid. 2016. "DFT Study of Adsorption of Diazinon, Hinosan, Chlorpyrifos and Parathion Pesticides on the Surface of B36N36 Nanocage and Its Fe Doped Derivatives as New Adsorbents," *Acta Physico-Chimica Sinica* 32, 5:1191-1198 doi: 10.3866/PKU.WHXB201603021.

[31] Klamt, Andreas. J. 1995. "Conductor-like Screening Model for Real Solvents: A New Approach to the Quantitative Calculation of Solvation Phenomena" *J. Phys. Chem.* 99, 7: 2224-2235 https://doi.org/10.1021/j100007a062.

INDEX

D

E

F

G

H

I

K

L

M

N

O

P

Q

R

S

T

W